T0180965

Sound Scattering on Spherical Objects

Tom Rother

Sound Scattering
on Spherical Objects

 Springer

Tom Rother
German Aerospace Center
Neustrelitz, Mecklenburg-Vorpommern,
Germany

ISBN 978-3-030-36450-2   ISBN 978-3-030-36448-9   (eBook)
https://doi.org/10.1007/978-3-030-36448-9

This Springer imprint is published by the registered company Springer Nature Switzerland AG
The registered company address is: Gewerbestrasse 11, 6330 Cham, Switzerland

Preface

Scattering of scalar and vector wave fields is a basic physical process that is of some importance for non-destructive diagnostic purposes in many technical and medical fields, among others. One need only think of radar and sonar techniques for target detection, remote sensing of the atmosphere, ultrasonic measurements, the detection of suspensions in colloidal solutions by nephelometer measurements, lidar technologies, or electron scattering in material sciences, to mention only a few applications. Accordingly, there exist a wide range of textbooks for specialists which require already a corresponding professional experience. On the other hand, there is to state a lack of an introductory literature that covers basic aspects of modeling such scattering processes on a level of upper undergraduate or graduate students, let's say. Filling this gap is within the heart of this book. To this end, the focus is put on acoustic plane wave scattering on different spherical objects. Acoustic plane waves, because this results in dealing with the simpler situation of scalar wave fields. And the restriction to spherical objects is due to the fact that we can operate on a sound mathematical basis regarding the existence of solutions and the convergence behavior. But it is another advantage of this restriction that this geometry results in further simplifications. This will allow the reader to better follow the details of the basic concepts and methods discussed in this book. But it will be sketched out at appropriate places how these methods can be generalized to become applicable to more complex scattering situations, and where the interested reader can find corresponding references. The book at hand is moreover written from a pragmatic rather than a stringent mathematical point of view. This is due to the fact that in most situations of practical interest one is operating in the gray zone between mathematical stringency and practicability. Today, there exist a variety of freely available but sophisticated computer programs to analyze complex scattering processes. However, to find out if a rigorous numerical solution is a physically reasonable one that provides a sufficient level of accuracy, or to find out the limit of a certain approximation is often not even a simple task. A better understanding of what happens behind the scene can be helpful in such situations. It is therefore another goal of this book to provide appropriate criteria to estimate the reliability and accuracy of obtained scattering results. Each chapter of this book is therefore equipped with a collection of related Python programs which can freely be modified, combined, and/or matched to the reader's needs. All these programs are also contained in the Extra Materials which appear together with the book. There one can also find a data base with nearly 60,000 precalculated scattering cross-sections for different bisphere configurations. This data base can be used for intercomparison purposes, practical applications, and as a starting point for a data base with an even finer resolution. It can moreover be used with benefit to discuss averaged scattering quantities, as it will be done in the last section of this book.

What awaits the reader when reading the book? The first chapter serves as an introduction to the underlying physical and mathematical aspects, equations, and tools to formulate and solve the scattering problems covered by this book. This is followed by a chapter that is concerned with the calculation of the so-called "T-matrices" of single-homogeneous spheres, and of concentric two-layered spheres with both these objects centered in the laboratory frame. The laboratory frame denotes the coordinate system at which the scattering process takes place. The T-matrix method—although not the only possibility—is a powerful tool for solving scattering problems not only on spherical but also on nonspherical objects, and to model multiple scattering. Since directly linked to the Green's functions, the T-matrices are appropriate to study basic mathematical and physical properties of scattering processes. They provide moreover a link to other solution methods like surface integral equation techniques and point-matching methods. The scattering behavior of a special spherical object—the so-called "Janus sphere"—is studied in the third chapter of this book. Janus spheres have a quite interesting potential for new applications especially in nanotechnologies, and more and more literature on this aspect can be found. This book ends with a chapter regarding sound scattering on bispheres. This is especially aimed at introducing the basic concept of solving multiple scattering processes if an ensemble of scatterer has to be taken into account.

I would like to express my appreciation to the German Aerospace Center for the space of autonomy and the administrative support it gives me to write this book. Dr. Zachary Evenson and Elke Sauer from Springer deserve a word of thanks for their interest and continuous support during the publication process.

Neustrelitz, Germany   Tom Rother
Autumn 2019

Contents

Chapter 1
Basics of Plane Wave Scattering

Since we are interested in acoustic plane wave scattering on spherical objects, the scalar Helmholtz equation in spherical coordinates represents the starting point for all the considerations within this book. The relevant eigensolutions of this equation and their properties are within the focus of this chapter. They are used to represent all the involved fields in terms of a series expansion (a Fourier series). Taking the radiation condition at infinity and appropriate boundary conditions at the spherical surface of the scatterer into account, we know from mathematics that the solution of such a scattering problem exists, and, moreover, that the representation in terms of a Fourier series converges against this unique solution. The "appropriate boundary conditions" are the homogeneous Dirichlet and Neumann condition in the case of a sound soft and sound hard sphere, two continuity conditions in the case of a sound penetrable sphere, and the limiting situation of a so-called Robin boundary condition in the case of Janus spheres. The sound mathematical basis regarding the uniqueness and convergence of the solution gets lost if nonspherical scatterers are considered. The reader who may be interested in such mathematical aspects is referred to [1–3], for example. Here we will trust the mathematicians and start directly with the formulation of the scattering problem without any proofs. In acoustics we are always concerned with the solution of only one scalar scattering problem. But at the end of this chapter we will give a perspective of how to use this knowledge to solve the scattering problem of electromagnetic waves on spherical objects by employing the so-called Debye potentials. This reduces the vector scattering problem to the solution of two scalar problems.

A scattering experiment measures certain quantities in the far-field of the scatterer. The far-field is of some importance since every measurement outside the direction of propagation of the primary incident plane wave and far away from the scattering object can solely be related to the wave scattered by this object (see [4], Chap. 4

Electronic supplementary material The online version of this chapter (https://doi.org/10.1007/978-3-030-36448-9_1) contains supplementary material, which is available to authorized users.

© Springer Nature Switzerland AG 2020
T. Rother, *Sound Scattering on Spherical Objects*,
https://doi.org/10.1007/978-3-030-36448-9_1

therein). The measurement quantities of our interest throughout this book are the differential and total scattering cross-sections. Both these quantities are determined from the intensity of the scattered field in the far-field. These are not the only but often used quantities measured in a scattering experiment.

1.1 Helmholtz Equation and Eigensolutions

A typical scattering configuration is shown in Fig. 1.1. A sphere with radius $r = a$ is centered in the laboratory frame. The primary incident plane wave u_{inc} is traveling along the positive z-axis and produces the scattered field u_s outside the sphere. In case of a sound penetrable sphere there will also exist an internal field u_{int} inside the sphere that must be taken into account to solve the scattering problem. The scattering plane is represented by the x-z-plane of the laboratory frame with θ denoting the scattering angle. This is the general scattering configuration we have in mind throughout this book. The only thing that changes is the scattering object itself.

The basic equation for any of the fields involved in the scattering process is given by

$$\left(\nabla^2 + k^2\right) u(r, \theta, \phi) \; = \; 0 \; . \tag{1.1}$$

This is the scalar Helmholtz equation that looks quite simple as long as the Laplace operator ∇^2 remains a pure symbolic quantity. But in spherical coordinates this operator is given by

$$\nabla^2 \; = \; \frac{1}{r^2} \frac{\partial}{\partial r} \left(r^2 \frac{\partial}{\partial r} \right) + \frac{1}{r^2 \sin \theta} \frac{\partial}{\partial \theta} \left(\sin \theta \frac{\partial}{\partial \theta} \right) + \frac{1}{r^2 \sin^2 \theta} \frac{\partial^2}{\partial \phi^2} \; . \tag{1.2}$$

This looks not that simple. But do not give up. We promise that you will become acquainted with its solutions after a while. $k = 2\pi/\lambda$ in this equation denotes the wave number that depends on the wavelength of the incident field and the properties of the material inside and outside the sphere. Outside the sphere we will always assume

Fig. 1.1 Scattering of a plane wave on a spherical object with radius $r = a$ centered in the laboratory frame. θ denotes the scattering angle in the scattering plane (the x-z-plane)

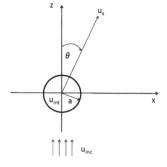

Table 1.1 Relations between the unit vectors in Cartesian and spherical coordinates

	$\hat{x}$	$\hat{y}$	$\hat{z}$
$\hat{r}$	$\sin\theta\cos\phi$	$\sin\theta\sin\phi$	$\cos\theta$
$\hat{\theta}$	$\cos\theta\cos\phi$	$\cos\theta\sin\phi$	$-\sin\theta$
$\hat{\phi}$	$-\sin\phi$	$\cos\phi$	0

vacuum represented by k_0. The wave number inside a sound penetrable sphere will be denoted with k_p.

The relations between the unit vectors in Cartesian and spherical coordinates are given in Table 1.1. $\theta \in [0, \pi]$ and $\phi \in [0, 2\pi]$ holds for the elevation and azimuthal angle. To be a little bit more general at this point, let us assume that we have a certain nonspherical scatterer with its boundary surface $\partial\Gamma$ given by the parameter representation

$$\vec{r} = r(\theta, \phi) \cdot \hat{r} \,. \tag{1.3}$$

Here are a few examples of rotational symmetric scatterers: The most simple geometry is, of course, the spherical surface given by

$$r(\theta) = a = const. \,. \tag{1.4}$$

A spheroidal particle with the z-axis being identical with its axis of revolution can be described by

$$r(\theta) = a \cdot \left[\cos^2\theta + \left(\frac{a}{b}\right)^2 \cdot \sin^2\theta\right]^{-1/2} \,. \tag{1.5}$$

"a" denotes the semi-axis along the axis of revolution, and "b" denotes the second semi-axis. $a/b < 1$ and $a/b > 1$ are the aspect ratios of oblate and prolate spheroids, respectively. $a/b = 1$ is again the sphere with radius $r = a$. Chebyshev particles are another kind of particles which are of some interest in scattering theory.

$$r(\theta) = a \cdot (1 + \epsilon \cdot \cos n\theta) \tag{1.6}$$

is the corresponding parameter representation of its boundary surface. "a" denotes the radius of the underlying sphere, "ϵ" is the deformation parameter, and "n" represents the order of the Chebyshev polynomial. The limiting case of a spherical particle with radius $r = a$ results obviously from $\epsilon = 0$. Two examples of such Chebyshev particles are shown in Fig. 1.2. There it was again assumed that the z-axis is the axis of revolution. Please, note that we use millimeter as the unit of length throughout this book. But this is not really a restriction since, regarding scattering we will see later on that it is characterized by the dimensionless size parameter $\beta = ka$. I.e., two scatterers with different radii but the same morphology and size parameter will show the same scattering behavior.

Fig. 1.2 Surface plot of two types of Chebyshev particles with $a = 3.0\,$mm, $\epsilon = 0.1$. Order of Chebyshev polynomial: $n = 3$ (**a**), $n = 5$ (**b**)

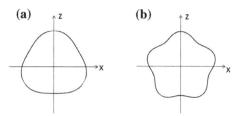

A surface element on the surface of the scatterer is given by

$$\mathrm{d}S = \left| \frac{\partial \vec{r}}{\partial \theta} \times \frac{\partial \vec{r}}{\partial \phi} \right| \mathrm{d}\theta \, \mathrm{d}\phi \,, \tag{1.7}$$

where

$$\frac{\partial \vec{r}}{\partial \theta} \times \frac{\partial \vec{r}}{\partial \phi} = r^2 \sin\theta \cdot \hat{r} - r \sin\theta \cdot \frac{\partial r}{\partial \theta} \cdot \hat{\theta} - r \cdot \frac{\partial r}{\partial \phi} \cdot \hat{\phi} \,. \tag{1.8}$$

The unit vector $\hat{n}$ at the boundary surface $\partial\Gamma$ pointing into the outer region is calculated according to

$$\hat{n} = \frac{\frac{\partial \vec{r}}{\partial \theta} \times \frac{\partial \vec{r}}{\partial \phi}}{\left| \frac{\partial \vec{r}}{\partial \theta} \times \frac{\partial \vec{r}}{\partial \phi} \right|} \,. \tag{1.9}$$

These relations are needed if one is interested not only in spherical but also in non-spherical scatterers. However, in the special case of the spherical boundary surface (1.4) we have simply

$$\mathrm{d}S = a^2 \sin\theta \, \mathrm{d}\theta \, \mathrm{d}\phi \,, \tag{1.10}$$

and $\hat{n}$ becomes identical with $\hat{r}$. This will make our live much more easy as we will see in the next chapter.

The two independent eigensolutions of (1.1) which are of importance for the scattering analysis are

$$\psi_{ln}(kr, \theta, \phi) = j_n(kr) \cdot Y_{ln}(\theta, \phi) \tag{1.11}$$
$$\varphi_{ln}(kr, \theta, \phi) = h_n^{(1)}(kr) \cdot Y_{ln}(\theta, \phi) \,. \tag{1.12}$$

$Y_{ln}(\theta, \phi)$ are the spherical harmonics

$$Y_{ln}(\theta, \phi) = \sqrt{\frac{2n + 1}{4\pi} \frac{(n - l)!}{(n + l)!}} \cdot P_n^l(\cos\theta) \, e^{il\phi} \,. \tag{1.13}$$

These functions are normalized to unity, i.e., they fulfill the orthonormality relation

$$\int_0^{2\pi} d\phi \int_0^{\pi} d\theta \sin\theta \, Y^*_{l'n'}(\theta, \phi) \, Y_{ln}(\theta, \phi) = \delta_{ll'} \, \delta_{nn'} \, . \tag{1.14}$$

j_n in (1.11) and $h_n^{(1)}$ in (1.12) are the spherical Bessel and Hankel functions of first kind. The spherical Hankel function of first kind is the sum of the spherical Bessel and Neumann function $y_n(kr)$,

$$h_n^{(1)}(kr) = j_n(kr) + i \cdot y_n(kr) \, , \tag{1.15}$$

where i denotes the imaginary unit. A closer look at the asymptotic behavior

$$\lim_{r \to \infty} j_n(kr) = \frac{1}{kr} \cdot \cos\left[kr - \frac{\pi}{2}(n+1)\right] \tag{1.16}$$

$$\lim_{r \to \infty} h_n^{(1)}(kr) = (-i)^{n+1} \cdot \frac{e^{ikr}}{kr} \tag{1.17}$$

shows that only $h_n^{(1)}$ is in agreement with the radiation condition

$$\lim_{r \to \infty} \left(\frac{\partial}{\partial r} - ik\right) u_s(r, \theta, \phi) = 0 \left(\frac{1}{r}\right) \, . \tag{1.18}$$

This condition must hold for the scattered field far away from the scatterer (i.e., in the far-field) and independent of its geometry, as already mentioned at the beginning of this chapter. $0(\frac{1}{r})$ in (1.18) means that all terms with a higher power of $1/r$ are neglected. While $j_n(kr)$ is a regular function everywhere in space, $h_n^{(1)}(kr)$ becomes singular if r tends to zero. The functions $P_n^l(\cos\theta)$ in (1.13) are the associated Legendre polynomials. They may be calculated from the conventional Legendre polynomials $P_n(x)$ by use of the relation

$$P_n^l(x) = (-1)^l \cdot (1 - x^2)^{l/2} \cdot \frac{d^l P_n(x)}{dx^l} \tag{1.19}$$

with the $P_n(x)$ given by

$$P_n(x) = \frac{1}{2^n n!} \left(\frac{d}{dx}\right)^n (x^2 - 1)^n \, . \tag{1.20}$$

$$P_n^{-l}(\cos\theta) = (-1)^l \cdot \frac{(n-l)!}{(n+l)!} \cdot P_n^l(\cos\theta) \tag{1.21}$$

is the relation between the associated Legendre polynomials with positive and negative index l.

These are the functions we will frequently employ in what follows. They have quite interesting features and obey interesting relations which are of importance in scattering theory and discussed in several textbooks (see the appendices in [5,

6], for example). Here we will simply consider these functions to be given by the corresponding routines of the special functions module of SciPy—a scientific Python package. These are:

- Spherical harmonics:

scipy.special.sph_harm(m,n,theta,phi)

Please, note that "m" in the list of arguments corresponds to "l" in (1.13), and that the meaning of "theta" and "phi" in the list of arguments is interchanged with their meaning in (1.13) (i.e., θ is the azimuthal angle running from 0 to 2π, and ϕ is the elevation angle running from 0 to π in the special functions module of SciPy)!!!

- Associated Legendre polynomials:

scipy.special.lpmn(m,n,z)

"z" is the real-valued argument of the Legendre polynomial.

- spherical Bessel's function:

scipy.special._jn(n,z,derivative=False)

"z" is the in general complex-valued argument of the Bessel function. If "derivative=True" is used instead, the derivative of this function with respect to its argument is calculated.

- spherical Neumann's function:

scipy.special._yn(n,z,derivative=False)

"z" is the in general complex-valued argument of the Neumann function. If "derivative=True" is used instead, the derivative of this function with respect to its argument is calculated.

For more details see the reference guide of SciPy. These routines are used in the corresponding Python programs of this book. Using Python and Matplotlib it is quite easy to write a short plot program to become acquainted with the behavior of these functions. An example of such a program for the spherical Bessel functions (program *plot_Bessel_function.py*) is provided with the Python routines of this chapter.

1.2 Transformation Behavior of the Eigensolutions

The transformation behavior of the eigenfunctions (1.11) and (1.12) with respect to a translation and a rotation of the laboratory frame is of central importance not only for all the scattering configurations considered in this book but also for scattering on nonspherical objects centered in the laboratory frame. The latter situation is met, for example, if the symmetry axis of a rotational symmetric but nonspherical scatterer

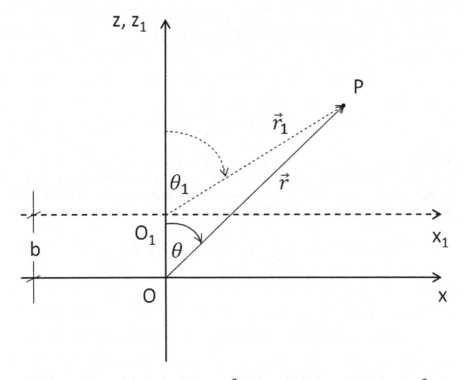

Fig. 1.3 Translation of the laboratory frame by $\vec{b} = b\hat{z}$ along the positive z-axis. $\vec{r} = \vec{r}_1 + \vec{b}$ holds between the position vectors of the laboratory frame $\{x, y, z\}$ and the shifted frame $\{x_1, y_1, z_1\}$

does not agree with the z-axis of the laboratory frame. To benefit from the symmetry properties of its boundary surface we have to rotate the laboratory frame appropriately (see [3, 5], for example). A detailed mathematical analysis of the transformation properties and their importance for multiple scattering can be found in [7]. I highly recommend this book for those readers who are interested in such mathematical aspects. However, while reading this book I missed the practical and computational aspects. To at least partially add these aspects was another incentive to write the book at hand.

1.2.1 Translation Along the Z-Axis

The most simple translation is a translation along the z-axis of the laboratory frame, as depicted in Fig. 1.3. Since both systems show the same dependence on the azimuthal angle ϕ, we have only to consider the following transformation between the (r, θ)-coordinates of system $\{x, y, z\}$ and the (r_1, θ_1)-coordinates of system $\{x_1, y_1, z_1\}$:

$$r_1 = \left(b^2 + r^2 - 2br \cdot \cos\theta\right)^{1/2}, \tag{1.22}$$

$$\theta_1 = \pi - \arccos\left(\frac{r^2 - r_1^2 - b^2}{-2r_1 b}\right) , \tag{1.23}$$

$$r = \left(b^2 + r_1^2 - 2br_1 \cdot \cos\theta_1\right)^{1/2} , \tag{1.24}$$

and

$$\theta = \arccos\left(\frac{r_1^2 - r^2 - b^2}{-2rb}\right) . \tag{1.25}$$

We are now interested in expressing the eigensolutions $\psi_{l,n}(r, \theta, \phi)$ and $\varphi_{l,n}(r, \theta, \phi)$ in a certain point of observation of the laboratory frame by the corresponding eigensolutions $\psi_{l,n}(r_1, \theta_1, \phi_1)$ and $\varphi_{l,n}(r_1, \theta_1, \phi_1)$ of the translated system. This can be accomplished as follows (see [7], Chap. 3):

$$\psi_{ln}(kr, \theta, \phi) = \sum_{\nu=0}^{\nu cut}(-1)^n \cdot \widehat{S}_{n\nu}^l(kb) \cdot \psi_{l\nu}(kr_1, \theta_1, \phi_1) , \tag{1.26}$$

and

$$\varphi_{ln}(kr, \theta, \phi) = \sum_{\nu=0}^{\nu cut}(-1)^n \cdot S_{n\nu}^l(kb) \cdot \psi_{l\nu}(kr_1, \theta_1, \phi_1) , \quad r_1 < b \tag{1.27}$$

$$\varphi_{ln}(kr, \theta, \phi) = \sum_{\nu=0}^{\nu cut}(-1)^n \cdot \widehat{S}_{n\nu}^l(kb) \cdot \varphi_{l\nu}(kr_1, \theta_1, \phi_1) , \quad r_1 > b . \tag{1.28}$$

$\widehat{S}_{n\nu}^l(kb)$ and $S_{n\nu}^l(kb)$ are the separation matrices for which we have the following relations:

$$\widehat{S}_{n\nu}^l(kb) = Re\left\{S_{n\nu}^l(kb)\right\} , \tag{1.29}$$

$$S_{n\nu}^{ll'}(kb) = (-1)^{n+\nu} \cdot S_{n\nu}^{ll'}(-kb) = (-1)^n \cdot S_{n\nu}^l(kb) \cdot \delta_{ll'} , \tag{1.30}$$

and

$$S_{n\nu}^l(kb) = S_{\nu n}^l(kb) = S_{n\nu}^{-l}(kb) . \tag{1.31}$$

$Re\{\cdots\}$ in (1.29) denotes the real part of the quantity in the brackets. Using (1.30) we can perform a translation along the negative z-axis of the laboratory frame. And we can also see that the mode l remains unchanged by this transformation. The separation matrix $S_{n\nu}^{ll'}(\vec{b})$ of an arbitrary translation $\vec{b}$ is related to the so-called Gaunt coefficients. These coefficients can alternatively be used to perform a translation of the laboratory frame. But in this book we choose another approach that is based on a combination of a translation along an appropriate z-axis and one or two rotations. This is due to the fact that the translation along the z-axis results in a quite simple separation matrix. Different expressions for this special separation matrix are known. The following expression was derived by Martin:

$$S_{n\nu}^l(kb) = (-1)^l \cdot i^{n+\nu} \cdot \frac{e^{ikb}}{ikb} \cdot \sum_{j=|l|}^{n+\nu} j! \cdot \left(\frac{i}{2kb}\right)^j \cdot \sum_{s=s_0}^{s_1} A_s(\nu, |l|) \cdot A_{j-s}(n, -|l|),$$

(1.32)

where

$$s_0 = \max(0, j - n)$$

(1.33)

$$s_1 = \min(\nu, j - |l|)$$

(1.34)

and

$$A_\alpha(n, l) = \sqrt{2n+1} \cdot \sqrt{\frac{(n+l)!}{(n-l)!}} \cdot \frac{(n+\alpha)!}{(l+\alpha)! \cdot \alpha! \cdot (n-\alpha)!}.$$

(1.35)

Martin could show that his expression is mathematically equivalent to the expression derived by Rehr, Albers, and Fritzsche that is also frequently used in the literature. However, the expression derived by Martin has several advantages from a numerical point of view. It is therefore used throughout this book (for more details on this aspect, see [7], Chap. 3.16). The *basics.py* module that comes along with this book contains a corresponding Python program. This program can be used to test the relations (1.30) and (1.31), for example.

The Python program *trans_single_wf_co.py* in the collection of programs that belongs to this chapter allows the test of the translation behavior of the eigensolutions with azimuthal mode $l = 0$ according to (1.26)–(1.28), and for a dimensionless shift kb of the laboratory frame along its z-axis. The eigensolutions in this program are at first calculated for observation points with a fixed distance kr but at equidistant angles $\theta \in [0, \pi]$ in the laboratory frame. The obtained values are compared afterwards with the values calculated from the corresponding representation in terms of the eigensolutions of the shifted system with coordinates of the observation points according to (1.22) and (1.23). Eigenfunctions with mode $l = 0$ are the only eigenfunctions we have to take into account in scattering configurations with only a shift of the laboratory frame along its z-axis, as we will see later on. However, it is not too difficult with the Python functions listed above to modify this program to become applicable also to eigenfunctions with modes $l \neq 0$. But, now, let us have a look at a few examples. We want to test the transformation of the regular and outgoing eigensolutions $\psi_{01}(kr = 1.0)$ and $\varphi_{01}(kr = 1.0)$ given in the laboratory frame if this frame is shifted afterwards by $kb = 2.0$ along its positive z-axis. The computation in the laboratory frame is performed in steps of $5°$ with respect to $\theta \in [0, \pi]$. In so doing, we have to distinguish the two cases $kr_1 < kb$ and $kr_1 > kb$ for the outgoing eigensolution. The results are shown in Figs. 1.4, 1.5, 1.6 and 1.7. Even for a small truncation parameter νcut the transformation works quite well for the regular eigensolution and the real part of the outgoing eigensolution, but it fails for its imaginary part. This happens especially for values with $kr_1 \approx kb$ (see Fig. 1.6). This situation improves only at higher truncation parameters. However the region $kr_1 \approx kb$ remains critical!

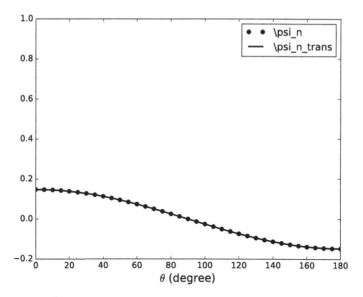

Fig. 1.4 Regular eigensolution ψ_{01} with $kr = 1.0$, $kb = 2.0$, and $\nu cut = 4$

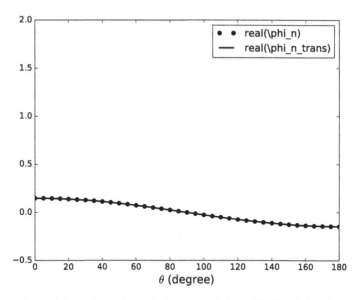

Fig. 1.5 Real part of the outgoing eigensolution φ_{01} with $kr = 1.0$, $kb = 2.0$ and $\nu cut = 4$

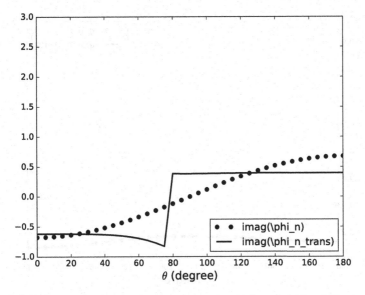

Fig. 1.6 Imaginary part of the outgoing eigensolution φ_{01} with $kr = 1.0, kb = 2.0$ and $\nu cut = 4$

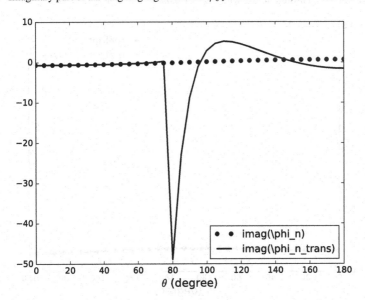

Fig. 1.7 Imaginary part of the outgoing eigensolution φ_{01} with $kr = 1.0, kb = 2.0$ and $\nu cut = 12$

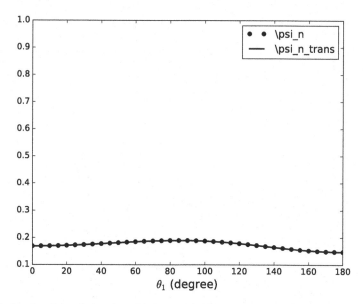

Fig. 1.8 Regular eigensolution ψ_{01} with $kr_1 = 1.0$, $kb = 2.0$, and $\nu cut = 4$

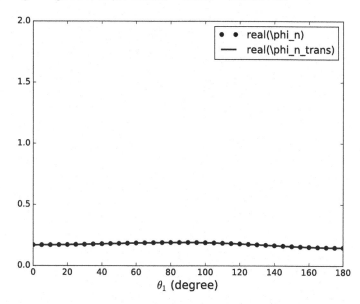

Fig. 1.9 Real part of the radiating eigensolution φ_{01} with $kr_1 = 1.0$, $kb = 2.0$ and $\nu cut = 4$

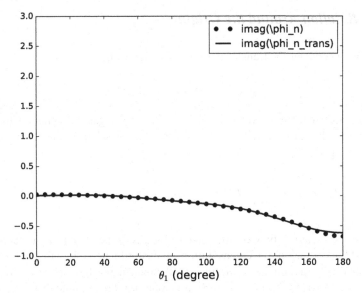

Fig. 1.10 Imaginary part of the radiating eigensolution φ_{01} with $kr_1 = 1.0, kb = 2.0$ and $\nu cut = 4$

Fortunately, we will never meet this situation in all the scattering configurations discussed in this book. The situation we are always faced with is reflected in the Python program *trans_single_wf_col.py*. It works as follows: The observation points are now given in the shifted system with fixed values of kr_1, and at different angles θ_1. The shift is chosen such that $kb \geq 2 \cdot kr_1$ holds. The equal sign holds if we consider the scattering behavior of touching bispheres as it is discussed in the last chapter of this book, or if a centered sphere is shifted by its diameter along the z-axis, for example. Regarding the bispheres, this is the most difficult situation since the interaction between the spheres is strongest in that case. But back to our Python program: In the next step the eigensolutions are calculated at these observation points but in the laboratory frame. They are finally transformed into the shifted frame. In so doing, we have to take only the two relations (1.26) and (1.27) into account. The results of such a transformation for ψ_{01} and φ_{01} are presented in Figs. 1.8, 1.9 and 1.10, and if $kr_1 = 1.0$ and $kb = 2.0$ is chosen. $\theta_1 \in [0, \pi]$ was considered in steps of $5°$. Now we can state an excellent agreement for all quantities already at the lower truncation parameter. It should be mentioned that these tests are an appropriate way to estimate the range of applicability of a certain numerical implementation of the separation matrices with respect to the maximum shift and the maximum order of the eigensolutions in a scattering simulation.

1.2.2 Rotation Around the Origin

We will now consider the transformation of the eigensolutions

$$\xi_{ln}(kr, \theta, \phi) = \chi_n(kr) \cdot Y_{ln}(\theta, \phi) \tag{1.36}$$

under the rotation of the coordinate system. $Y_{ln}(\theta, \phi)$ are again the spherical harmonics given in (1.13), and $\chi_n(kr)$ denotes the spherical Bessel and Hankel functions of first kind, respectively.

$$\xi_{ln}(kr, \theta, \phi) = \sum_{l_1=-n}^{n} D_{l_1 l}^{(n)}(-\gamma, -\theta_p, -\alpha) \cdot \xi_{l_1 n}(kr_1, \theta_1, \phi_1) \tag{1.37}$$

establishes the relation between the eigensolutions $\xi_{ln}(kr, \theta, \phi)$ of the laboratory frame and the corresponding eigensolutions of a coordinate system that results from a rotation of the laboratory frame in a mathematical positive sense by the three Eulerian angles $(\alpha, \theta_p, \gamma)$ (see [5, 7, 8], for example). Please, note that we used the Eulerian angle θ_p instead of its commonly used notation β for not to be confused with the size parameter β that appears also frequently in our equations. The first rotation is a rotation by an angle of $\alpha \in [0, 2\pi]$ around the original z-axis. The second rotation is a rotation by an angle of $\theta_p \in [0, \pi]$ around the new y-axis after the first rotation. And the final rotation is a rotation by an angle of $\gamma \in [0, 2\pi]$ around the new z-axis after the second rotation. To find out the meaning of "in mathematical positive sense", please check the web. There are a lot of corresponding animations.

If we want to express the eigensolutions of the rotated coordinate system in terms of the eigensolutions of the laboratory frame we have on the other hand

$$\xi_{l_1 n}(kr_1, \theta_1, \phi_1) = \sum_{l=-n}^{n} D_{l l_1}^{(n)}(\alpha, \theta_p, \gamma) \cdot \xi_{ln}(kr, \theta, \phi) . \tag{1.38}$$

Only the azimuthal modes l or l_1 are obviously affected by these transformations. The quantity $D_{l_1 l}^{(n)}(\alpha, \theta_p, \gamma)$ denotes the matrix of rotation. It is given by

$$D_{l l'}^{(n)}(\alpha, \theta_p, \gamma) = e^{-il\alpha} \cdot d_{l l'}^{(n)}(\theta_p) \cdot e^{-il'\gamma} , \tag{1.39}$$

where

$$d_{l l'}^{(n)}(\theta_p) = \sum_{s=s_{\min}}^{s_{\max}} \frac{\left[(n+l)!(n-l)!(n+l')!(n-l')!\right]^{1/2}}{s!(l+l'+s)!(n-l-s)!(n-l'-s)!} \times$$

$$(-1)^{n+l'+s} \left[\sin\frac{\theta_p}{2}\right]^{2n-2s-l-l'} \left[\cos\frac{\theta_p}{2}\right]^{2s+l+l'} \tag{1.40}$$

are the Wigner d-functions, and

$$s_{\min} = \max[0, -(l + l')] \tag{1.41}$$

$$s_{\max} = \min(n - l, n - l') . \tag{1.42}$$

Matrix $\mathbf{D}^{(n)}(\alpha, \theta_p, \gamma)$ is a unitary matrix and obeys the relations

$$\left[\mathbf{D}^{(n)}(\alpha, \theta_p, \gamma)\right]^{-1} = \mathbf{D}^{(n)}(-\gamma, -\theta_p, -\alpha) , \tag{1.43}$$

and

$$D_{ll'}^{(n)}(0, 0, 0) = \delta_{ll'} . \tag{1.44}$$

This matrix is also implemented in the Python module *basics.py* that comes along with this book. All the scattering configurations we are faced with in this book are characterized by a rotational symmetry. Regarding the matrix of rotation, we can therefore generally choose $\gamma = 0°$.

1.2.3 Two Combinations

We are now prepared to look at two combinations of the rotation and the shift along a z-axis we will employ in the following chapters. The first one is depicted in Fig. 1.11 and runs as follows: First we rotate the laboratory frame $\{x, y, z\}$ into the system $\{x', y', z'\}$ by use of the two Eulerian angles α and θ_p. Relation (1.37) can then be used to express the eigensolutions of the laboratory frame in terms of the eigensolutions of the rotated system. In a second and final step we perform the shift by b along the new z'-axis and apply relations (1.26)–(1.28) for the two eigensolutions of our interest in the rotated system. Introducing the two quantities (let us call them operators with the subindex "rs" denoting a rotation and a shift!)

$$\left[\widehat{T}_{\mathrm{rs}}\right]_{nn'}^{ll'} = (-1)^n \cdot D_{l'l}^{(n)}(0, -\theta_p, -\alpha) \cdot \widehat{S}_{nn'}^{l'}(kb) \tag{1.45}$$

and

$$\left[T_{\mathrm{rs}}\right]_{nn'}^{ll'} = (-1)^n \cdot D_{l'l}^{(n)}(0, -\theta_p, -\alpha) \cdot S_{nn'}^{l'}(kb) \tag{1.46}$$

we end up with the following representation of the eigensolutions given in the laboratory frame in terms of the eigenfunctions given in the rotated and shifted system $\{x'', y'', z''\}$:

$$\psi_{ln}(kr, \theta, \phi) = \sum_{l'=-n}^{n} \sum_{n'=0}^{n'cut} \left[\widehat{T}_{\mathrm{rs}}\right]_{nn'}^{ll'} \cdot \psi_{l'n'}(kr'', \theta'', \phi'') \tag{1.47}$$

Fig. 1.11 Combination of a rotation and a shift along the new z'-axis after the rotation

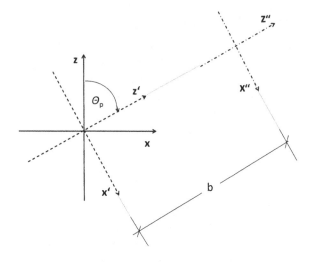

$$\varphi_{ln}(kr,\theta,\phi) = \sum_{l'=-n}^{n} \sum_{n'=0}^{n'cut} [T_{rs}]_{nn'}^{ll'} \cdot \psi_{l'n'}(kr'',\theta'',\phi'') ; \quad r'' < b \qquad (1.48)$$

$$\varphi_{ln}(kr,\theta,\phi) = \sum_{l'=-n}^{n} \sum_{n'=0}^{n'cut} [\widehat{T}_{rs}]_{nn'}^{ll'} \cdot \varphi_{l'n'}(kr'',\theta'',\phi'') ; \quad r'' > b . \qquad (1.49)$$

The inverse transformation, i.e., expressing the eigensolutions in the rotated and shifted system in terms of the eigensolutions of the laboratory frame can be obtained from the same steps as above but in reverse order. First we have to accomplish the shift by $-b$ along the z'-axis by taking relation (1.30) into account. And finally we have to reverse the initial rotation back to the laboratory frame by using (1.38). In so doing, we get

$$\psi_{ln}(kr'',\theta'',\phi'') = \sum_{n'=0}^{n'cut} \sum_{l'=-n'}^{n'} [\widehat{T}_{rs}^{-1}]_{nn'}^{ll'} \cdot \psi_{l'n'}(kr,\theta,\phi) \qquad (1.50)$$

$$\varphi_{ln}(kr'',\theta'',\phi'') = \sum_{n'=0}^{n'cut} \sum_{l'=-n'}^{n'} [T_{rs}^{-1}]_{nn'}^{ll'} \cdot \psi_{l'n'}(kr,\theta,\phi) ; \quad r < b \qquad (1.51)$$

$$\varphi_{ln}(kr'',\theta'',\phi'') = \sum_{n'=0}^{n'cut} \sum_{l'=-n'}^{n'} [\widehat{T}_{rs}^{-1}]_{nn'}^{ll'} \cdot \varphi_{l'n'}(kr,\theta,\phi) ; \quad r > b , \qquad (1.52)$$

where

$$[\widehat{T}_{rs}^{-1}]_{nn'}^{ll'} = (-1)^{n'} \cdot \widehat{S}_{nn'}^{l}(kb) \cdot D_{l'l}^{(n')}(\alpha,\theta_p,0) \qquad (1.53)$$

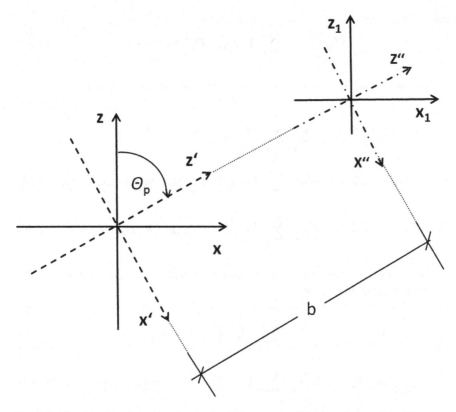

Fig. 1.12 Combination of two rotations and a shift to accomplish a translation of the laboratory frame $\{x, y, z\}$ into the system $\{x_1, y_1, z_1\}$

and

$$\left[T_{rs}^{-1}\right]_{nn'}^{ll'} = (-1)^{n'} \cdot S_{nn'}^{l}(kb) \cdot D_{l'l}^{(n')}(\alpha, \theta_p, 0) . \tag{1.54}$$

The second combination we will employ in this book is presented in Fig. 1.12. It is a translation of the laboratory frame, accomplished by an additional rotation compared to the combination discussed above. I.e. all the axes of the final system $\{x_1, y_1, z_1\}$ are now in parallel to the axes of the laboratory frame. This additional rotation is therefore simply the reverse of the first rotation of the former combination. Now we have the two operators (with the subindex "rsr" denoting a rotation, a subsequent shift, and the final rotation!)

$$\left[\widehat{T}_{rsr}\right]_{nn'}^{l\,l_1} = \sum_{l'=-n}^{n} \left[\widehat{T}_{rs}\right]_{nn'}^{l\,l'} \cdot D_{l_1 l'}^{(n')}(\alpha, \theta_p, 0) \tag{1.55}$$

and

$$[T_{\text{rsr}}]_{nn'}^{l l_1} = \sum_{l'=-n}^{n} [T_{\text{rs}}]_{nn'}^{l l'} \cdot D_{l_1 l'}^{(n')}(\alpha, \theta_p, 0) . \tag{1.56}$$

The eigensolutions given in the laboratory frame can now be expressed according to

$$\psi_{ln}(kr, \theta, \phi) = \sum_{n'=0}^{n'cut} \sum_{l_1=-n'}^{n'} \left[\widehat{T}_{\text{rsr}}\right]_{nn'}^{l l_1} \cdot \psi_{l_1 n'}(kr_1, \theta_1, \phi_1) \tag{1.57}$$

$$\varphi_{ln}(kr, \theta, \phi) = \sum_{n'=0}^{n'cut} \sum_{l_1=-n'}^{n'} [T_{\text{rsr}}]_{nn'}^{l l_1} \cdot \psi_{l_1 n'}(kr_1, \theta_1, \phi_1) ; \ r_1 < b \tag{1.58}$$

$$\varphi_{ln}(kr, \theta, \phi) = \sum_{n'=0}^{n'cut} \sum_{l_1=-n'}^{n'} \left[\widehat{T}_{\text{rsr}}\right]_{nn'}^{l l_1} \cdot \varphi_{l_1 n'}(kr_1, \theta_1, \phi_1) ; \ r_1 > b \tag{1.59}$$

in terms of the eigenfunctions of the translated system $\{x_1, y_1, z_1\}$. And, finally, the inverse transformation reads

$$\psi_{ln}(kr_1, \theta_1, \phi_1) = \sum_{n'=0}^{n'cut} \sum_{l_1=-n'}^{n'} \left[\widehat{T}_{\text{rsr}}^{-1}\right]_{nn'}^{l l_1} \cdot \psi_{l_1 n'}(kr, \theta, \phi) \tag{1.60}$$

$$\varphi_{ln}(kr_1, \theta_1, \phi_1) = \sum_{n'=0}^{n'cut} \sum_{l_1=-n'}^{n'} \left[T_{\text{rsr}}^{-1}\right]_{nn'}^{l l_1} \cdot \psi_{l_1 n'}(kr, \theta, \phi) ; \ r < b \tag{1.61}$$

$$\varphi_{ln}(kr_1, \theta_1, \phi_1) = \sum_{n'=0}^{n'cut} \sum_{l_1=-n'}^{n'} \left[\widehat{T}_{\text{rsr}}^{-1}\right]_{nn'}^{l l_1} \cdot \varphi_{l_1 n'}(kr, \theta, \phi) ; \ r > b , \tag{1.62}$$

where

$$\left[\widehat{T}_{\text{rsr}}^{-1}\right]_{nn'}^{l l_1} = \sum_{l'=-n}^{n} D_{l' l}^{(n)}(0, -\theta_p, -\alpha) \cdot \left[\widehat{T}_{\text{rs}}^{-1}\right]_{nn'}^{l' l_1} \tag{1.63}$$

and

$$\left[T_{\text{rsr}}^{-1}\right]_{nn'}^{l l_1} = \sum_{l'=-n}^{n} D_{l' l}^{(n)}(0, -\theta_p, -\alpha) \cdot \left[T_{\text{rs}}^{-1}\right]_{nn'}^{l' l_1} . \tag{1.64}$$

Although these equations look quite complex due to the many summations and indices, it is not that complicate. I encourage the reader to test these transformations by writing his own program. This is a not too difficult exercise if using the corresponding subroutines of the *basics.py* module.

1.3 Series Expansions of the Fields and Scattering Quantities

1.3.1 Series Expansions

Regarding the scattering problems of our interest we are concerned with three different types of fields. These are the primary incident field u_{inc} that is considered to be given, the scattered field u_s outside the scatterer, and a possibly existing internal field u_{int} inside the scatterer. Both fields u_s and u_{int} result from the interaction of the scatterer with the primary incident field. To determine the scattered field of a scattering configuration with a given morphology of the scatterer, i.e., to describe the interaction of the primary incident field with the scatterer is the final goal of all our efforts. An overview of existing methods together with a discussion of their advantages and disadvantages can be found in [3, 5, 7], for example. The so-called "T-matrix method" is within the focus of this book. If the scatterer under consideration is that of a homogeneous sphere centered in the laboratory frame this method is identical with the well-known Mie theory (or Debye-Mie theory). To approximate all the involved fields by an appropriate series expansion in terms of the eigensolutions of the underlying Helmholtz equation (that is, by using a Fourier series) is within the heart of this approach. The following expansions are used for our purposes:

- The primary incident field u_{inc}:

$$u_{inc}(k_0 r, \theta, \phi) = \sum_{n=0}^{ncut} \sum_{l=-n}^{n} d_{ln} \cdot \psi_{ln}(k_0 r, \theta, \phi) \,, \qquad (1.65)$$

- The internal field u_{int}:

$$u_{int}(k_p r, \theta, \phi) = \sum_{n=0}^{ncut} \sum_{l=-n}^{n} g_{ln} \cdot \psi_{ln}(k_p r, \theta, \phi) \,, \qquad (1.66)$$

- The scattered field u_s:

$$u_s(k_0 r, \theta, \phi) = \sum_{n=0}^{ncut} \sum_{l=-n}^{n} c_{ln} \cdot \varphi_{ln}(k_0 r, \theta, \phi) \,. \qquad (1.67)$$

d_{ln}, g_{ln}, and c_{ln} are the so far unknown expansion coefficients, and k_0 and k_p denote the wave numbers outside and inside the scatterer. That the regular eigensolutions are used for the primary incident and internal field is a consequence of the regularity requirement for both these fields everywhere in space and at the origin of the coordinate system. Using the radiating eigensolutions for the scattered field is a consequence of the radiation condition (see the remark subsequent to (1.18)). From a stringent mathematical point of view these fields are in fact represented by an infinite

series expansion. However, in any numerical realization we have to live with a finite
series expansion, i.e., we have to choose the finite truncation parameter *ncut* appropriately. To find out what "appropriately" means for a given scattering configuration
can become a challenging task and depends not at least on the application one has
in mind. This aspect will therefore be discussed frequently in this book. It should
also be mentioned that the T-matrix approach used in this book and based upon the
above given expansions of the fields, can be applied not only to spherical coordinates.
Two-dimensional polar coordinates, for example, is another system that is frequently
used to study the scattering behavior of infinitely extended cylinders with circular or
noncircular cross-sections in different configurations. The corresponding mathematical background and the transformation behavior of the eigensolutions can be found
in [7]. But since the concepts and approaches considered in this book are applicable
to cylindrical coordinates as well, I decided to put the focus only on problems that
can be solved in spherical coordinates.

The plane wave

$$u_{\text{inc}}(k_0 r, \theta) = U_0 \cdot e^{ik_0 z} = U_0 \cdot e^{ik_0 r \cdot \cos\theta} \tag{1.68}$$

is used to represent the primary incident field in all the scattering configurations of
our interest. It travels along the positive z-direction of the laboratory frame. And since
any scattering intensities are calculated with respect to the intensity of the incident
field we can further choose $U_0 = 1$. Fortunately, a simple expansion in terms of the
regular eigensolutions is known for this plane wave. This expansion reads

$$u_{\text{inc}}(k_0 r, \theta) = \sum_{n=0}^{ncut} d_n \cdot \psi_n(k_0 r, \theta) \tag{1.69}$$

with the expansion coefficients d_n given by

$$d_n = \sqrt{4\pi(2n+1)} \cdot i^n . \tag{1.70}$$

Note that only the eigensolutions with the azimuthal mode $l = 0$ are needed in
this expansion. How does this expansion looks like if expressed in terms of the
regular eigensolutions of the rotated and/or shifted coordinate systems discussed in
the previous section? From (1.37) we get for a simple rotation

$$u_{\text{inc}}(k_0 \bar{r}, \bar{\theta}, \bar{\phi}) = \sum_{n=0}^{ncut} \sum_{\bar{l}=-n}^{n} \widehat{d}_{\bar{l}n} \cdot \psi_{\bar{l}n}(k_0 \bar{r}, \bar{\theta}, \bar{\phi}) , \tag{1.71}$$

where

$$\widehat{d}_{\bar{l}n} = D_{\bar{l}0}^{(n)}(0, -\theta_p, -\alpha) \cdot d_n . \tag{1.72}$$

This transformation is of our interest in the third chapter when discussing the scattering behavior of arbitrarily oriented Janus spheres centered in the laboratory frame. The transformation that is accomplished by a shift of the laboratory frame along its z-axis is just as simple. Taking (1.26) into account the expansion (1.69) of the plane wave is transformed into

$$u_{\text{inc}}(k_0 r_1, \theta_1, \phi_1) = \sum_{n'=0}^{n'cut} \widehat{d}_{n'} \cdot \psi_{n'}(k_0 r_1, \theta_1, \phi_1) \,, \tag{1.73}$$

where

$$\widehat{d}_{n'} = \sum_{n=0}^{ncut} (-1)^n \cdot \widehat{S}_{nn'}^0(k_0 b) \cdot d_n \,. \tag{1.74}$$

This transformation is required in the fourth chapter, for example, where we discuss the scattering behavior of bispheres aligned along the z-axis of the laboratory frame. For the two combinations discussed in the previous subsection we have on the other hand

$$u_{\text{inc}}(k_0 r'', \theta'', \phi'') = \sum_{l'=-n}^{n} \sum_{n'=0}^{n'cut} \widehat{d}_{l'n'} \cdot \psi_{l'n'}(k_0 r'', \theta'', \phi'') \,, \tag{1.75}$$

where

$$\widehat{d}_{l'n'} = \sum_{n=0}^{ncut} \left[\widehat{T}_{\text{rs}}\right]_{nn'}^{0l'} \cdot d_n \tag{1.76}$$

follows from (1.47), and

$$u_{\text{inc}}(k_0 r_1, \theta_1, \phi_1) = \sum_{n'=0}^{n'cut} \sum_{l_1=-n'}^{n'} \widehat{d}_{l_1 n'} \cdot \psi_{l_1 n'}(k_0 r_1, \theta_1, \phi_1) \,, \tag{1.77}$$

where

$$\widehat{d}_{l_1 n'} = \sum_{n=0}^{ncut} \left[\widehat{T}_{\text{rsr}}\right]_{nn'}^{0l_1} \cdot d_n \tag{1.78}$$

results from (1.57). Using (1.43) it is not too difficult to show that (1.73)/(1.74) is the limiting result of the last two transformations if $\alpha = \theta_p = 0°$. All these expressions are given here to demonstrate that, given the incident plane wave in the laboratory frame this wave is also known in all the other coordinate systems of our interest. Especially (1.73)/(1.74) and (1.77)/(1.78) can moreover be used to test the correct numerical implementation of a certain transformation, as we will demonstrate now.

Expansion (1.69)/(1.70) of the incident plane wave (1.68), if considered in the laboratory frame, was transformed into a corresponding expansion in terms of the eigensolutions of the frame shifted along the z-axis by using (1.73)/(1.74). But regarding this special shift there exists a much easier way that was already used in [9]. We must simply replace the incident field $\exp(ik_0 z)$ that holds in the laboratory frame by the expression $\exp(ik_0(z_1 + b))$ that holds in the shifted frame. Separation of the term $\exp(ik_0 b)$ and expanding the remaining term $\exp(ik_0 z_1)$ according to (1.69) and (1.70) but now in the shifted frame results in the transformed expansion coefficients

$$\widehat{d}_{n'} = e^{ik_0 b} \cdot d_{n'} \tag{1.79}$$

as an alternative to (1.74), and with $d_{n'}$ according to (1.70). No separation matrix is obviously required in this case! But since both expressions must be identical

$$e^{ik_0 b} \cdot d_{n'} = \sum_{n=0}^{ncut} (-1)^n \cdot \widehat{S}_{n n'}^0 (k_0 b) \cdot d_n \tag{1.80}$$

follows in a straightforward way. This is a first sum rule that must hold for the separation matrix $\widehat{S}_{n n'}^0 (k_0 b)$.

Another sum rule follows from the fact that any arbitrary translation of the laboratory frame results in the analytical expression

$$u_{\text{inc}}(k_0 r_1, \theta_1) = e^{ik_0(b \cdot \cos \theta_p + r_1 \cdot \cos \theta_1)} \tag{1.81}$$

for the incident plane wave (1.68) if expressed in the coordinates of the translated system. First we note that this expression is again independent on the angle ϕ_1, as it happens already in the laboratory frame. And, second, this expression must be identical with the much more complicate expressions (1.77)/(1.78) where the matrix of rotation and the separation matrix is involved. The Python program *plane_wave_trans.py* of this chapter can be used to test the numerical equivalence of both expressions, and to determine the truncation parameter that is required to achieve a certain numerical accuracy with the transformation (1.77)/(1.78). More sum rules will be discussed later on in Chap. 4.

1.3.2 Scattering Quantities

In all the situations considered in this book the coefficients $c_{l n}$ of expansion (1.67) of the scattered field in the laboratory frame are given by the general relation

$$c_{l n} = \sum_{l' n'} [T]_{n n'}^{l l'} \cdot d_{l' n'} . \tag{1.82}$$

$[T]_{n n'}^{l l'}$ are the elements of the "T-matrix" (or "T-operator"), and the coefficients $d_{l' n'}$ are calculated from the given expansion coefficients (1.70) of the incident plane wave. Let us assume that we are already in the possession of both these quantities for a given scattering configuration. Using (1.82) in (1.67), and if taking the asymptotic behavior (1.17) in the far-field into account gives

$$u_s(k_0 r, \theta, \phi) = -\frac{i}{k_0} \cdot \sum_{n=0}^{ncut} \sum_{l=-n}^{n} \sum_{l' n'} (-i)^n \cdot [T]_{n n'}^{l l'} \cdot d_{l' n'} \cdot Y_{l n}(\theta, \phi) \cdot \frac{e^{i k_0 r}}{r} .$$

(1.83)

I.e., the scattered field is an outgoing spherical wave in the far-field with the scattering amplitude function

$$f(\theta, \phi) = -\frac{i}{k_0} \cdot \sum_{n=0}^{ncut} \sum_{l=-n}^{n} \sum_{l' n'} (-i)^n \cdot [T]_{n n'}^{l l'} \cdot d_{l' n'} \cdot Y_{l n}(\theta, \phi) . \qquad (1.84)$$

The differential scattering cross-section—denoted with $d\sigma_s/d\omega$—is defined as the square of the scattering amplitude function,

$$\frac{d\sigma_s}{d\Omega} := |f(\theta, \phi)|^2 = f^*(\theta, \phi) \cdot f(\theta, \phi) . \qquad (1.85)$$

It is proportional to the angular dependent intensity of the scattered field measured in a corresponding scattering experiment in the far-field. "$d\Omega = \sin \theta \, d\theta \, d\phi$" is the differential solid angle element. It should be emphasized that $d\sigma_s/d\Omega$ is only a formal notation of the differential scattering cross-section. It must not be confused with an actually derivation! This is the most important scattering quantity we will use throughout this book to characterize the scattering behavior of a certain configuration. It is the ECG of a scattering process, so to speak.

The other quantity of our interest—the total scattering cross-section σ_s—is obtained from the solid angle integration of the differential scattering cross-section,

$$\sigma_s := \oint \frac{d\sigma}{d\Omega} \, d\Omega = \int_0^{2\pi} d\phi \int_0^{\pi} d\theta \sin \theta \, f^*(\theta, \phi) \cdot f(\theta, \phi) . \qquad (1.86)$$

In so doing, and since $Y_{l n}(\theta, \phi)$ is the only θ- and ϕ-dependent quantity in expression (1.84) of $f(\theta, \phi)$, we can benefit from the orthonormality relation (1.14). However, there exists a much easier way to calculate the total scattering cross-section. The so-called "optical theorem" relates the imaginary part of the scattering amplitude in forward direction to the total scattering cross-section, i.e., it holds

$$\sigma_s = \frac{4\pi}{k_0} \cdot Im\{f(\theta = 0)\} . \qquad (1.87)$$

This interesting relation holds not only for acoustic but many other scattering processes, including electromagnetic and quantum mechanical scattering processes. It has a long and interesting history, and its general proof is not that simple (for more about that, see [3, 10, 11], for example). It should also be mentioned that relation

(1.87) is only true if there is no absorption inside the scattering particle. If absorption must be considered in a scattering process, then (1.87) holds only for the so-called extinction cross-section that describes the combined effect of scattering outside the particle and absorption inside the particle. However, since absorption is neglected in all our examples we can rely on (1.87). If there raises any doubt about the correctness of this relation one can at least numerically compare this result with the result that follows from (1.86). This intercomparison becomes especially simple if the T-matrix in (1.84) reduces to

$$[T]_{nn'}^{ll'} = \frac{1}{2} \cdot \left(e^{2i\Delta_n} - 1\right) \cdot \delta_{ll'} \cdot \delta_{nn'} , \tag{1.88}$$

where Δ_n is a real-valued quantity called the "phase shift", and if $\tilde{d}_{l'n'}$ reduces to d_n of (1.70). We will see in the next chapter that this holds for some of our basic scattering objects. Here, let us consider it as an assured fact. Since

$$\left(e^{2i\Delta_n} - 1\right) = 2i \sin \Delta_n \cdot e^{i\Delta_n} \tag{1.89}$$

the scattering amplitude (1.84) reads

$$f(\theta, \phi) = \frac{\sqrt{4\pi}}{k_0} \cdot \sum_{n=0}^{ncut} \sin \Delta_n \cdot e^{i\Delta_n} \cdot \sqrt{2n+1} \cdot Y_{0n}(\theta, \phi) . \tag{1.90}$$

Taking into account that

$$Y_{0n}(\theta = 0, \phi) = \sqrt{\frac{2n+1}{4\pi}} \tag{1.91}$$

holds, we get

$$f(\theta = 0, \phi) = \frac{1}{k_0} \cdot \sum_{n=0}^{ncut} \sin \Delta_n \cdot e^{i\Delta_n} \cdot (2n+1) \tag{1.92}$$

for the scattering amplitude in forward direction. Its imaginary part becomes therefore

$$Im\{f(\theta = 0)\} = \frac{1}{k_0} \cdot \sum_{n=0}^{ncut} (2n+1) \cdot \sin^2 \Delta_n . \tag{1.93}$$

On the other hand, if using (1.90) in (1.86), and if taking the orthogonality relation (1.14) of the spherical harmonics into account, we end up with

$$\sigma_s = \frac{4\pi}{k_0^2} \cdot \sum_{n=0}^{ncut} (2n+1) \cdot \sin^2 \Delta_n . \tag{1.94}$$

This proves (1.87) for this special situation. The optical theorem is a remarkable result since it relates the intensity of the scattered field if summed up over all directions to the phase shift between the primary incident plane wave and the scattered field in forward direction.

1.4 About the Accuracy of Scattering Solutions

To find out if a numerically obtained scattering solution is physical meaningful and accurate enough for the applicational purposes one has in mind can become a quite challenging task. It depends to a great extent on the scattering configuration itself, on the physical model used to match this configuration, on the applied mathematical method, and, last but not least, on the numerical implementation of that method. Performing an intercomparison with a corresponding scattering experiment is one way to gain trust into a numerically obtained solution. But in most cases of practical interest it is again not even a simple task to perform such an experiment and to provide results within satisfying error bars (see Chaps. 12 and 13 in [12], for example). Producing trustworthy numerical and/or experimental results often requires long-term experiences in both these fields.

Focusing on spherical scattering objects throughout this book results in several simplifications, as already mentioned in the Preface. Regarding the field expansions (1.65)–(1.67) we know from mathematics that the expansion functions (1.11) and (1.12) form an orthonormal basis on the spherical surface of the scatterer centered in the considered coordinate system, and if k^2 is not an eigenvalue of the interior Dirichlet problem. The same holds for the derivatives of these functions with respect to the radius if k^2 is not an eigenvalue of the interior Neumann problem. This is of some importance for us since we intend to apply appropriate boundary conditions on the spherical surface of the scatterer to determine the unknown expansion coefficients of the scattered field by using a least-squares approach. It is thus guaranteed that this solution will converge in a least-squares sense against the unique solution of the scattering problem everywhere outside the scatterer. Moreover, the thus determined expansion coefficients are final [1]. That is, the corresponding T-matrix becomes a diagonal matrix. We will come back to this aspect in the next chapter. Corresponding numerical results are often used not only as benchmarks for other solution methods but also in real experiments for calibration and validation purposes. These nice properties get lost, and we have to abandon the knowledge about the convergence if we want to apply the same method to a scatterer with a nonspherical boundary surface. This is indeed possible but makes accuracy considerations even more complicate (see [3], for example).

Benchmark results for centered spheres are also of importance for the bisphere configurations we intend to discuss in chapter four. Regarding bispheres, we are always concerned with one sphere shifted off-center in the laboratory frame. This forces us to employ the matrix of rotation and the separation matrix in the course of solving this scattering problem. Its solution depends therefore from the correct numerical implementation and the accuracy of the computation of these matrices. Beside using the sum rules discussed at the end of Sect. 1.3.2 corresponding tests can be accomplished in the following way: These matrices are not involved in the solution process for a certain sphere that is centered in the laboratory frame. On the other hand, we know from physics that any local shift of a scatterer in the laboratory frame produces only a pure phase term in the corresponding scattered field if it is considered in the far-field. But since the differential scattering cross-section is calculated from (1.85) this shift is washed out. The differential scattering cross-section must therefore be identical for both the centered and the shifted sphere. The same holds for the total scattering cross-section since the shift of the scattered field in forward direction is compensated by a corresponding shift of the incident field. An intercomparison of both results will allow us therefore to test the correct numerical implementation and the accurate computation of the matrix of rotation and the separation matrix. Examples are given in the next chapter.

To test the fulfilment of the so-called "reciprocity condition"—a condition that can be traced back to the symmetry of the Green's function that can be related to every scattering process (see [3, 13], for example)—provides another possibility to gain trust into an obtained scattering solution. Let us consider the following two bisphere configurations to demonstrate the meaning of reciprocity: The bispheres in both configurations are assumed to be still the same, i.e., in Fig. 1.13 (b) the bispheres of (a) are just rotated by an angle of 90° in the x-z-plane. We assume further that the sources of the incident plane wave u_{inc} and the observation point P_o are located in the far-field of the bispheres in both configurations. Then, reciprocity tells us that we have to measure the same differential scattering cross-section in observation point P_o of configuration (a) and (b). It should be mentioned that reciprocity is much more general and independent of the morphology of the scatterer and the location of the source and observation point. But the above considered restrictions are used for all our tests in the last two chapters.

Another possibility to test the correct numerical implementation of a certain method and its accuracy is the intercomparison of its results with an independent

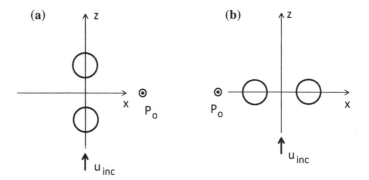

Fig. 1.13 Two bisphere configurations to test reciprocity

solution method, of course. This will be demonstrated in chapter four. There we will derive a general method that holds for arbitrarily oriented bispheres. But if this approach is applied to bispheres placed on the z-axis of the laboratory frame (see the configuration of Fig. 1.13a) it can be compared with the results of a much simpler and independent approach that allows us to solve the scattering problem for this specific bisphere configuration. We will show moreover that the lowest order iteration of this general method neglects the interaction between the component spheres. On the other hand, neglecting this interaction from the very beginning results again in a much simpler method where only the T-matrices of the component spheres and an analytical phase term get involved. The intercomparison of the results of both these independent approaches provides as again with some information about the accuracy of the obtained solutions and the correct numerical implementation of the decisive elements of the more general method. This is possible not least because of the fact that an increase of the distance between the component spheres in the general method results in a decrease of the interaction effect. And a final hint for those who want to start with developing an own scattering program or want to start using an existing software: Compare, compare, compare, ... !

1.5   Debye Potentials for Electromagnetic Scattering

This book focuses exclusively on acoustic wave scattering problems which can be traced back to the solution of the scalar Helmhotz equation, as outlined in the previous sections. Regarding the problem of electromagnetic wave scattering, for example, we have to fight in general with the more complex vector-wave equation

$$\left(\nabla \times \nabla \times - k^2\right) \vec{u}(r, \theta, \phi) = \vec{0} \tag{1.95}$$

that holds for the electric and magnetic field. Its relevant eigensolutions can be calculated in the following way from the eigensolutions (1.11)–(1.12) of the scalar Helmholtz equation:

$$\vec{\psi}_{l,n,1}(kr, \theta, \phi) = \frac{1}{\sqrt{n(n+1)}} \nabla \times \left(\hat{r} \cdot r \cdot \psi_{l,n}(kr, \theta, \phi)\right) \tag{1.96}$$

$$\vec{\psi}_{l,n,2}(kr, \theta, \phi) = \frac{1}{\sqrt{n(n+1)}} \frac{1}{k} \nabla \times \nabla \times \left(\hat{r} \cdot r \cdot \psi_{l,n}(kr, \theta, \phi)\right) \tag{1.97}$$

$$\vec{\varphi}_{l,n,1}(kr, \theta, \phi) = \frac{1}{\sqrt{n(n+1)}} \nabla \times \left(\hat{r} \cdot r \cdot \varphi_{l,n}(kr, \theta, \phi)\right) \tag{1.98}$$

$$\vec{\varphi}_{l,n,2}(kr, \theta, \phi) = \frac{1}{\sqrt{n(n+1)}} \frac{1}{k} \nabla \times \nabla \times \left(\hat{r} \cdot r \cdot \varphi_{l,n}(kr, \theta, \phi)\right) . \tag{1.99}$$

Corresponding transformation properties of a rotation and/or a shift of the coordinate system are also known for these eigensolutions (a comprehensive overview of these transformations can be found in the Appendices of [5]). Using these vectorial eigensolutions as expansion functions for the electromagnetic fields allows us therefore to apply the T-matrix approach that is used throughout this book also to electromagnetic scattering problems. This is described in detail in textbooks like [3, 5], for example, for a variety of different scattering geometries and configurations. However, focusing again on spherical scatterers results in a further simplification. Instead of solving the vector-wave equation we can solve two independent but scalar problems both related to the scalar Helmholtz equation. The relevant scalar quantities are the so-called Debye potentials Π_e and Π_m, named after Debye who invented these potentials for the first time to describe electromagnetic plane wave scattering on spherical obstacles [14]. I.e., instead of solving (1.95) we can solve

$$\nabla^2 \Pi_{e/m} + k^2 \Pi_{e/m} = 0 . \tag{1.100}$$

All the electromagnetic field components can be calculated from these potentials in the following way:

$$E_r(r, \theta, \phi) = \frac{i}{\omega\epsilon} \left[\frac{\partial^2(r \cdot \Pi_e(r, \theta, \phi))}{\partial r^2} + k^2 r \, \Pi_e(r, \theta, \phi) \right] \tag{1.101}$$

$$E_\theta(r, \theta, \phi) = \frac{i}{\omega\epsilon} \frac{1}{r} \cdot \frac{\partial^2(r \cdot \Pi_e(r, \theta, \phi))}{\partial r \partial \theta} + \frac{1}{\sin\theta} \cdot \frac{\partial \Pi_m(r, \theta, \phi)}{\partial \phi} \tag{1.102}$$

$$E_\phi(r, \theta, \phi) = \frac{i}{\omega\epsilon} \frac{1}{r \sin\theta} \cdot \frac{\partial^2(r \cdot \Pi_e(r, \theta, \phi))}{\partial r \partial \phi} - \frac{\partial \Pi_m(r, \theta, \phi)}{\partial \theta} , \tag{1.103}$$

and

$$H_r(r, \theta, \phi) = -\frac{i}{\omega\mu_0} \left[\frac{\partial^2(r \cdot \Pi_m(r, \theta, \phi))}{\partial r^2} + k^2 r \, \Pi_m(r, \theta, \phi) \right] \tag{1.104}$$

$$H_\theta(r, \theta, \phi) = -\frac{i}{\omega\mu_0} \frac{1}{r} \cdot \frac{\partial^2(r \cdot \Pi_m(r, \theta, \phi))}{\partial r \partial \theta} + \frac{1}{\sin\theta} \cdot \frac{\partial \Pi_e(r, \theta, \phi)}{\partial \phi} \tag{1.105}$$

$$H_\phi(r, \theta, \phi) = -\frac{i}{\omega\mu_0} \frac{1}{r \sin\theta} \cdot \frac{\partial^2(r \cdot \Pi_m(r, \theta, \phi))}{\partial r \partial \phi} - \frac{\partial \Pi_e(r, \theta, \phi)}{\partial \theta} . \tag{1.106}$$

Regarding a plane electric field

$$\vec{E}_{\text{inc}}(r, \theta, \phi) = \hat{x} \cdot E_0 \cdot e^{ik_0 r \cos\theta} \tag{1.107}$$

that is polarized in x-direction and travels along the positive z-axis of the laboratory frame we know the following series expansions for the related Debye potentials:

$$\Pi_e^{\text{inc}}(r, \theta, \phi) = \frac{k_0}{\omega\mu_0} \cdot E_0 \cdot \cos\phi \cdot \sum_{n=1}^{\infty} i^n \frac{(2n+1)}{n(n+1)} \cdot j_n(k_0 r) \cdot P_n^1(\cos\theta) , \tag{1.108}$$

and

$$\Pi_m^{\text{inc}}(r, \theta, \phi) = - E_0 \cdot \sin \phi \cdot \sum_{n=1}^{\infty} i^n \frac{(2n+1)}{n(n+1)} \cdot j_n(k_0 r) \cdot P_n^1(\cos \theta) . \quad (1.109)$$

The derivation of these expressions can be found in many textbooks on Mie theory, like in the famous book [15] (Chap. 13.5 therein), for example. For the corresponding potentials of the scattered and internal field components we can therefore choose the *ansatz*

$$\Pi_e^s(r, \theta, \phi) = \frac{k_0}{\omega \mu_0} \cdot E_0 \cdot \cos \phi \cdot \sum_{n=1}^{\infty} a_n \cdot h_n^{(1)}(k_0 r) \cdot P_n^1(\cos \theta) \quad (1.110)$$

$$\Pi_m^s(r, \theta, \phi) = - E_0 \cdot \sin \phi \cdot \sum_{n=1}^{\infty} b_n \cdot h_n^{(1)}(k_0 r) \cdot P_n^1(\cos \theta) \quad (1.111)$$

$$\Pi_e^{\text{int}}(r, \theta, \phi) = \frac{k}{\omega \mu_0} \cdot E_0 \cdot \cos \phi \cdot \sum_{n=1}^{\infty} c_n \cdot j_n(k r) \cdot P_n^1(\cos \theta) \quad (1.112)$$

$$\Pi_m^{\text{int}}(r, \theta, \phi) = - E_0 \cdot \sin \phi \cdot \sum_{n=1}^{\infty} d_n \cdot j_n(k r) \cdot P_n^1(\cos \theta) \quad (1.113)$$

with the so far unknown expansion coefficients a_n, b_n, c_n, and d_n. It is again the essential point to find appropriate solutions for these coefficients. And for spherical scattering geometries this is indeed as simple as it will be described in the next chapter for acoustic scattering since the required boundary conditions for the fields at the spherical surface can be reduced to corresponding boundary conditions for the Debye potentials. An application of this approach to the electromagnetic scattering problem on bispheres was first published in [16]. And it is also applicable to the scattering problem of electromagnetic waves on Janus spheres. Unfortunately, this nice possibility gets lost if dealing with electromagnetic scattering on nonspherical objects.

1.6 Python Programs

This section itemizes all the Python programs that will allow the reader to test the transformation properties of the eigenfunctions discussed in this chapter, to test the sum rule (1.80), and to test the translation behavior of the primary incident plane wave. It contains moreover the module "basics.py" that is required to run all the other Python programs which are provided with this book. All the necessary tools are gathered together in this module. But I should also emphasize at this place that I am not a software engineer, and that I started using Python very recently. I am therefore convinced that (and I would be happy if) a more skilled programmer will find ways of reorganizing the software to become faster and more efficient. However, it was my intention to apply not too sophisticated programming techniques so that the programs are formulated close to the mathematical expressions derived and discussed in this book. To run the programs you have to install Python (I used Python 3.4.4), SciPy (v. 0.18.0), NumPy (v. 1.11.0), and Matplotlib (v. 1.5.1). All the programs can also be downloaded from Springer's "extra materials" web page. But, now, let us take a short look at each single program.

- modul *basics.py*:
  As mentioned above, this module contains all the basic elements to perform the scattering analysis on the objects and configurations considered in this book. These are in particular the separation matrix and the matrix of rotation, all the T-matrices for the centered component spheres we will derive in the subsequent chapters, and the elements of the block matrix or the block operator that are used for the iterative solution of the T-matrix or T-operator equations for axisymmetrically and arbitrarily oriented bispheres. This module must always be in the actual working directory!

- program *plot_Bessel_function.py*:
  This is a simple plot program to plot the spherical Bessel's function of a given order. It uses the SciPy routine mentioned at the end of Sect. 1.1. Figure 1.14 shows the "matplotlib" output for $j_3(z)$ with $z \in [0, 15]$.

- programs *trans_single_wf_co.py* and *trans_single_wf_co1.py*:
  These two programs can be used to test the transformation of the regular and radiating eigensolution according to (1.29)–(1.31) but for two different situations. Both these situations are described in Sect. 1.2.1. "trans_single_wf_co.py" was used to produce the Figs. 1.4, 1.5, 1.6 and 1.7. Figures 1.8, 1.9 and 1.10, in contrast, have been generated with program "trans_single_wf_co1.py" that avoids situations where we have to distinguish between $r_1 < b$ and $r_1 > b$.

- program *sumtest.py*:
  This program provides a first possibility to test the correct numerical implementation and the range of applicability of the real part of the separation matrix (1.32) for the azimuthal mode $l = 0$ by using (1.80). It provides only numbers.

- program *plane_wave_trans.py*:
  This program provides a further test for the correct numerical implementation and for the range of applicability of the separation matrix and the matrix of rotation that is no longer restricted to the azimuthal mode $l = 0$. It considers an arbitrary translation of the primary incident plane wave in the laboratory frame if this wave is calculated in the local frame of the translated system and along the surface of a sphere with radius a_1 in steps of $10°$. It provides again only numbers for a better intercomparison. The first number in each line corresponds to the analytical expression (real and imaginary part) given by (1.81). The second number in each line corresponds to (1.77)/(1.78). Please, note that all length units are considered to be given in millimeter. But this is not a restriction since any other choice can be

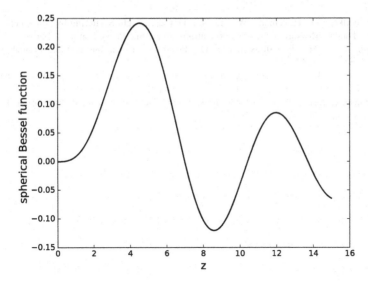

Fig. 1.14 Spherical Bessel's function $j_3(z)$

controlled by the wave number k since only the dimensionless expressions ka or kr appear in the equations and are of importance for the scattering process. The translation is described by the two Eulerian angles α and θ_p, and by the shift b of the coordinate system.

The full programs are given in Appendix A.

References

1. Sommerfeld, A.: Partial Differential Equations in Physics. Academic Press, New York (1949)
2. Colton, D., Kress, R.: Integral Equation Methods in Scattering Theory. Wiley, New York (1983)
3. Rother, T., Kahnert, M.: Electromagnetic Wave Scattering on Nonspherical Particles: Basic Methodology and Applications. Springer, Heidelberg (2014)
4. van de Hulst, H.C.: Light Scattering by Small Particles. Dover, New York (1981)
5. Mishchenko, M.I., Travis, L.D., Lacis, A.A.: Scattering, Absorption, and Emission of Light by Small Particles. Cambridge University Press, Cambridge (UK) (2002)
6. Abramowitz, M., Stegun, I.A.: Handbook of Mathematical Functions. Harri Deutsch, Frankfurt/Main (1984)
7. Martin, P.A.: Multiple Scattering: Interaction of Time-Harmonic Waves with N Obstacles. Cambridge University Press, Cambridge (UK) (2006)
8. Varshalovich, D.A., Moskalev, A.N., Khersonskii, V.K.: Quantum Theory of Angular Momentum. World Scientific, Singapore (1988)
9. Eyges, L.: Some nonseparable boundary value problems and the many-body problem. Ann. Phys. **2**, 101–128 (1957)
10. Newton, R.G.: Optical theorem and beyond. Am. J. Phys. **44**, 639–642 (1976)
11. Saxon, D.S.: Tensor scattering matrix for the electromagnetic field. Phys. Rev. **100**, 1771–1775 (1955)

12. Mishchenko, M.I., Hovenier, J.W., Travis, L.D. (eds.): Light Scattering by Nonspherical Particles: Theory, Measurement, and Applications. Academic Press, London (2000)
13. Rother, T.: Green's Functions in Classical Physics. Springer International Publishing AG, Cham, Switzerland (2017)
14. Debye, P.: Der Lichtdruck auf Kugeln von beliebigem Material. Ann. Phys. **30**, 57–136 (1909)
15. Born, M., Wolf, E.: Principles of Optics. Pergamon Press, Oxford (1980)
16. Borghese, F., Denti, P., Toscano, G., Sindoni, O.I.: Electromagnetic scattering by a cluster of spheres. Appl. Opt. **18**, 116–120 (1979)

Chapter 2
Scattering on Single Homogeneous and Two-Layered Spheres

After we have gathered the relevant eigensolutions of the scalar Helmholtz equation as well as some of their important properties in the previous chapter we are now prepared to determine the expansion coefficients of the scattered field if represented in terms of the radiating eigensolutions according to (1.67). Deriving the corresponding T-matrix is within the heart of this effort. The scattering problem is solved once this quantity is known. To apply this approach to different types of spherical scatterers will keep us busy throughout the remaining chapters. We start in this chapter with the most simple 3d-scattering objects, with three different types of homogeneous spheres and two different types of concentric two-layered spheres. It is first assumed that all these spheres are centered in the laboratory frame. But it is demonstrated in the third section that the obtained results for the centered spheres provide a further possibility to test the correct numerical implementation and the range of applicability of the separation matrix and the matrix of rotation. Both these matrices are the decisive elements in the T-matrix equations derived and solved in the subsequent chapters for the more complex scattering objects. This chapter ends again with a short description of the corresponding Python programs.

2.1 T-Matrices of Homogeneous Spheres

2.1.1 Sound Soft Sphere

The different types of spherical scatterers considered in this chapter are characterized by the different boundary conditions the involved fields have to fulfill along the surface of the sphere with radius $r = a$. Regarding the sound soft sphere, the homogeneous Dirichlet condition

---

Electronic supplementary material The online version of this chapter (https://doi.org/10.1007/978-3-030-36448-9_2) contains supplementary material, which is available to authorized users.

© Springer Nature Switzerland AG 2020
T. Rother, *Sound Scattering on Spherical Objects*,
https://doi.org/10.1007/978-3-030-36448-9_2

$$u_{\text{inc}}(\beta, \theta) + u_s(\beta, \theta) = 0 \tag{2.1}$$

for the sum of the primary incident plane wave and the scattered field must hold. This condition, together with the radiation condition (1.18) that is already taken into account in the expansion functions we used for the scattered field, ensures the unique solvability of this scattering problem from the very beginning—a gentle resting pillow for a physicist that can be lost quickly if he is interested in more complex scattering geometries (see [1] for more details on this mathematical aspect, for example). It is now a relatively simple task to derive a relation between the so far unknown expansion coefficients of the scattered field and the known expansion coefficients of the primary incident plane wave. To this end we have to insert both expansions (1.67) and (1.69) into (2.1). After multiplication of the resulting equation with $\sin \theta \cdot Y_{l',n'}^*(\theta, \phi)$, and after integration from 0 to π with respect to θ and from 0 to 2π with respect to ϕ we get

$$c_{ln} = -\frac{j_n(\beta)}{h_n^{(1)}(\beta)} \cdot \delta_{l0} \cdot d_n \tag{2.2}$$

if taking the orthonormality relation (1.14) of the spherical harmonics into account. Note that we have already introduced the important size parameter

$$\beta = k_0 \cdot a = 2\pi \frac{a}{\lambda} \tag{2.3}$$

that expresses the ratio of the radius of the sphere and the wavelength of the incident plane wave. This parameter allows a scaling of scattering processes with respect to these two quantities if the morphology of the scatterer does not change. Instead of (2.2) we may also write

$$c_{ln} = \sum_{n'=0}^{ncut} \sum_{l'=-n'}^{n'} [T_s]_{nn'}^{ll'} \cdot d_{l'n'} \tag{2.4}$$

with the thus introduced T-matrix

$$[T_s]_{nn'}^{ll'} := -\frac{j_n(\beta)}{h_n^{(1)}(\beta)} \cdot \delta_{ll'} \cdot \delta_{nn'} \tag{2.5}$$

for the sound soft sphere. Please, note that the known expression (1.70) for the expansion coefficients of the primary incident field has been reformulated as follows:

$$d_{l'n'} = \sqrt{4\pi(2n'+1)} \cdot i^{n'} \cdot \delta_{0l'} . \tag{2.6}$$

First, we notice that only the azimuthal mode with $l = 0$ must be considered for the scattered field, which already applies to the incident field. Second, we see that the coefficients c_{ln} are final, as mentioned in Sect. 1.4. That is, once we have determined a

certain number of expansion coefficients $c_{00}, \ldots, c_{0\,ncut}$, and if our accuracy requirement makes it necessary to take some more coefficients into account, then we have
to calculate only these additional coefficients. The coefficients we have determined
before remain unchanged. This property follows also from the diagonal character of
the T-matrix (2.5).

Equation (2.4), that agrees with (1.82), can be considered as a relation that
describes the Transition from the known expansion coefficients of the primary incident field to the expansion coefficients of the scattered field we are looking for. This
is the origin of the notation "T-matrix"—the shortform of transition matrix. However, there exists an alternative interpretation of the T-matrix that emphasizes its
Transformation character (it can sometimes be quite beneficial to look at the same
thing from different point of views). To show this, we have to remember the property
mentioned in Sect. 1.4, i.e., the property that both sets of expansion functions (1.11)
and (1.12) form an orthonormal basis on the spherical surface of the scatterer. We
start from the known expansion

$$u_{\text{inc}}(\beta, \theta) = \sum_{n=0}^{ncut} d_n \cdot \psi_n(\beta, \theta) \tag{2.7}$$

of the incident plane wave that holds on the spherical surface, and with coefficients
d_n given by (1.70). Then, we may ask how one can change from this representation
to the new representation

$$u_{\text{inc}}(\beta, \theta) = \sum_{n=0}^{ncut} \widehat{d}_n \cdot \varphi_n(\beta, \theta) \tag{2.8}$$

in terms of the radiating expansion functions but with the so far unknown coefficients
$\widehat{d}_n$. These coefficients can be determined if expressing the old basis functions $\psi_n(\beta, \theta)$
in terms of the new basis functions $\varphi_n(\beta, \theta)$. This is accomplished by introducing
the transformation matrix $\widehat{T}$ according to

$$\psi_n(\beta, \theta) = \sum_{n'=0}^{ncut} \left[\widehat{T}\right]_{n\,n'} \cdot \varphi_{n'}(\beta, \theta) \tag{2.9}$$

Multiplication of this equation with $\sin\theta \cdot Y_{0,n'}^*(\theta, \phi)$, and integration from 0 to π
with respect to θ results finally in

$$\left[\widehat{T}\right]_{n\,n'} = \frac{j_n(\beta)}{h_n^{(1)}(\beta)} \cdot \delta_{n\,n'} \tag{2.10}$$

for the elements of the transition matrix. It differs only in sign from (2.5)! That is,
the new expansion coefficients

$$\widehat{d}_n = \sum_{n'=0}^{ncut} [\widehat{T}]_{n\,n'} \cdot d_{n'} \tag{2.11}$$

in expansion (2.8) of the incident field, and with $d_{n'}$ given by (1.70), differ only in sign from the expansion coefficients $c_{0,n}$ we obtained for the scattered field! It should be mentioned that this transformation character of the T-matrix is not restricted to spherical scattering objects only, and that it can be used with benefit to prove some of the fundamental properties of the T-matrix. More informations on these aspects can be found in [2], Chap. 2 therein.

At the end of Sect. 1.3.2 we provided a proof of the correctness of the optical theorem to calculate the total scattering cross-section from the scattering amplitude function in forward direction. Relation (1.88) with a real-valued phase shift Δ_n was a precondition of that proof. Using (2.5) we are now in the possession of a concrete expression for the T-matrix of a sound soft sphere that allows us the proof of the correctness of this precondition, as promised in Sect. 1.3.2. Since the T-matrix (2.5) is a diagonal matrix it is sufficient for that proof to put the focus on its diagonal elements

$$[T_s]_{nn} = -\frac{j_n(\beta)}{h_n^{(1)}(\beta)} . \tag{2.12}$$

Let us "inflate" this expression as follows:

$$[T_s]_{nn} = \frac{1}{2} \cdot \left[\frac{-2j_n(\beta) + h_n^{(1)}(\beta) - h_n^{(1)}(\beta)}{h_n^{(1)}(\beta)} \right] . \tag{2.13}$$

Taking relation (1.15) into account it is straightforward to show that we get

$$[T_s]_{nn} = \frac{1}{2} \cdot ([S_s]_{nn} - 1) \tag{2.14}$$

with the elements

$$[S_s]_{nn} = -\frac{j_n(\beta) - i \cdot y_n(\beta)}{j_n(\beta) + i \cdot y_n(\beta)} = \frac{y_n(\beta) + i \cdot j_n(\beta)}{y_n(\beta) - i \cdot j_n(\beta)} . \tag{2.15}$$

of the thus introduced matrix $\mathbf{S_s}$. This is another famous matrix in scattering theory—the so-called "S-matrix"—that was independently introduced by Heisenberg and Wheeler to describe scattering processes in quantum mechanics. Looking at (2.14) the relation between this matrix and the T-matrix is given by

$$\mathbf{S_s} = 2\,\mathbf{T_s} + \mathbf{E} , \tag{2.16}$$

where $\mathbf{E}$ represents the unit matrix. Please, note that we will use boldface capital letters to denote matrices. Their elements are given by the corresponding capital letter in square brackets with upper and/or lower indices. Next, we have to show that the elements $[S_s]_{nn}$ can be expressed by

$$[S_s]_{nn} = e^{2i\Delta_n} \qquad (2.17)$$

with Δ_n representing a pure real-valued quantity. This is accomplished by using the identity

$$e^{2i\Delta_n} = \frac{e^{i\Delta_n}}{e^{-i\Delta_n}} = \frac{1 + i \cdot \tan\Delta_n}{1 - i \cdot \tan\Delta_n} . \qquad (2.18)$$

From (2.15) we thus get

$$\tan\Delta_n = \frac{j_n(\beta)}{y_n(\beta)} \qquad (2.19)$$

which is a real-valued quantity as long as β is a real-valued quantity. This proves relation (1.88) for the sound soft sphere! (2.17) proves moreover the unitarity property

$$\mathbf{S_s}^\dagger \cdot \mathbf{S_s} = \mathbf{E} \qquad (2.20)$$

of the S-matrix for a real-valued β. This is of some importance since the property of unitarity can be linked directly to the conservation of energy in scattering processes.

We will now show a few examples for three different size parameters of a sound soft sphere centered in the laboratory frame. The computations have been performed with the program *centered_sphere.py* that comes along with this chapter (see the description in the final section). The results are presented in Figs. 2.1, 2.2 and 2.3. We can observe more and more ripples in the side scattering region for an increasing size parameter. However, the highest intensity results always for the forward direction.

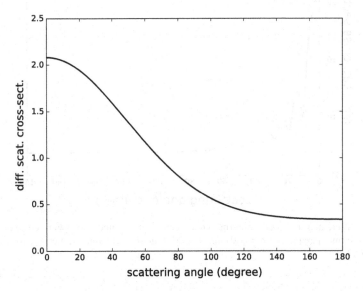

Fig. 2.1 Differential and total scattering cross-section of a sound soft sphere centered in the laboratory frame. Parameters: $a = 1.0$ mm, $\beta = 1.0$, *ncut* = 6. Total scattering cross-section: $\sigma_{tot} = 10.626$

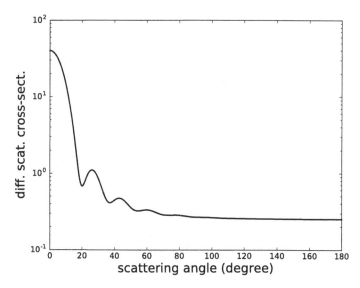

Fig. 2.2 Differential and total scattering cross-section of a sound soft sphere centered in the laboratory frame. Parameters: $a = 1.0\,\text{mm}$, $\beta = 10.0$, $ncut = 15$. Total scattering cross-section: $\sigma_{\text{tot}} = 7.531$

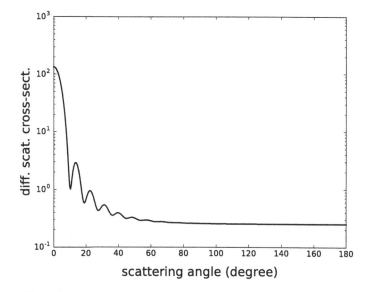

Fig. 2.3 Differential and total scattering cross-section of a sound soft sphere centered in the laboratory frame. Parameters: $a = 1.0\,\text{mm}$, $\beta = 20.0$, $ncut = 25$. Total scattering cross-section: $\sigma_{\text{tot}} = 7.092$

Table 2.1 Total scattering cross-section of a sound soft sphere with radius $a = 1.0\,\text{mm}$. It approaches 2π for increasing size parameters

β	σ_{tot}
1	10.626
50	6.732
100	6.569
150	6.502
200	6.464
500	6.381

Another important behavior can be observed if looking at the total scattering cross-section for increasing size parameters up to very large values. This quantity should approach the cross-sectional area $A = \pi a^2$ for the sphere with radius $r = a$, as one may expect from the geometric optics approximation (GO). But the computation reveals that it approaches twice this value, as demonstrated with Table 2.1 for a sphere with radius $a = 1.0\,\text{mm}$. This behavior is well-known and called the "extinction paradox". The general validity of this paradox for sound soft spheres at very large size parameters was demonstrated in a breathtaking analysis by Nussenzveig [3]. The reason for this discrepancy between the GO and the rigorous solution derived above is the fact that the GO, that is assumed to be applicable especially for very large size parameters, neglects diffraction. But this wave effect is enclosed in the rigorous solution. It is all the more remarkable that this paradox is not known to some authors. Although correct, they considered the obtained results for the total cross-section as erroneous, caused by mistakes in their computer code (see [4], for example). On the other hand, the GO approximation produces quite accurate results in the side- and backscattering region even at moderate size parameters, as one can see from Fig. 2.3 (see the region between 90° and 180° in the scattering plane). For more details on this aspect, see also [5], Chaps. 2 and 4 therein.

2.1.2 Sound Hard Sphere

The boundary condition

$$\frac{\partial}{\partial r}\left[u_{\text{inc}}(k_0 r, \theta) + u_{\text{s}}(k_0 r, \theta)\right]_{r=a} = 0 \tag{2.21}$$

for the sum of the incident and scattered field characterizes the sound hard sphere. To determine the unknown expansion coefficients c_{ln} of the scattered field we can follow exactly the same track used for the sound soft sphere. But since taking (2.21) instead of (2.1) into account we now end up with the T-matrix

$$[T_h]_{nn'}^{ll'} = -\frac{j_n'(\beta)}{h_n'^{(1)}(\beta)} \cdot \delta_{ll'} \cdot \delta_{nn'}, \tag{2.22}$$

while relation (2.4) remains the same. Please note that the dashes on the Bessel and Hankel functions denote the first derivatives with respect to their arguments! Since the coefficients (2.6) of the incident field are still the same we have again to consider only the azimuthal mode $l = 0$. The corresponding elements of the S-matrix read

$$[S_h]_{nn} = -\frac{j_n'(\beta) - i \cdot y_n'(\beta)}{j_n'(\beta) + i \cdot y_n'(\beta)} = \frac{y_n'(\beta) + i \cdot j_n'(\beta)}{y_n'(\beta) - i \cdot j_n'(\beta)}, \tag{2.23}$$

and we get

$$\tan \Delta_n = \frac{j_n'(\beta)}{y_n'(\beta)} \tag{2.24}$$

for the phase shift. This is again a real-valued quantity, and, consequently, S_h is again a unitary matrix as long as the size parameter β is real-valued.

We can immediately start to compare the results of sound hard spheres with the results we obtained before for the sound soft spheres. Figures 2.4, 2.5 and 2.6 show the differential and total scattering cross-sections for the same size parameters used for the sound soft spheres. Significant differences can be observed: Regarding the size parameter of $\beta = 1$, the backscattering intensity is now stronger than the intensity in forward direction. The total cross-sections are always below the corresponding cross-sections we have obtained for the sound soft spheres, and the value $2\pi a^2$ is now approached from below for increasing size parameters (see Table 2.2). And, finally, the ripples are more pronounced and extended up to the backscattering direction for

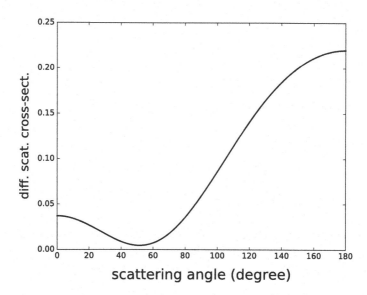

Fig. 2.4 Differential and total scattering cross-section of a sound hard sphere centered in the laboratory frame. Parameters: $a = 1.0\,\text{mm}$, $\beta = 1.0$, $ncut = 6$. Total scattering cross-section: $\sigma_{\text{tot}} = 1.010$

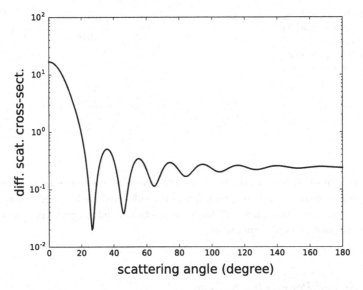

Fig. 2.5 Differential and total scattering cross-section of a sound hard sphere centered in the laboratory frame. Parameters: $a = 1.0\,$mm, $\beta = 10.0$, *ncut* = 15. Total scattering cross-section: $\sigma_{\text{tot}} = 4.930$

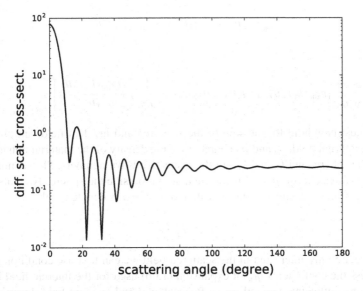

Fig. 2.6 Differential and total scattering cross-section of a sound hard sphere centered in the laboratory frame. Parameters: $a = 1.0\,$mm, $\beta = 20.0$, *ncut* = 25. Total scattering cross-section: $\sigma_{\text{tot}} = 5.458$

Table 2.2 Total scattering cross-section of a sound hard sphere with radius $a = 1.0\,$mm. It approaches 2π for increasing size parameters

β	σ_{tot}
1	1.010
50	5.853
100	6.020
150	6.083
200	6.118
500	6.193

the two size parameters $\beta = 10, 20$. This serves as a hint that the GO approximation, in contrast to sound soft spheres, may fail in the side- and backscattering region for such moderate size parameters. All these computations have again been performed with the program *centered_sphere.py*.

2.1.3 Sound Penetrable Sphere

The sound penetrable sphere is characterized by the two transmission conditions

$$u_{\text{inc}}(\beta, \theta) + u_s(\beta, \theta) = u_{\text{int}}(\beta_p, \theta) \tag{2.25}$$

and

$$\frac{1}{\rho_0} \frac{\partial}{\partial r} \left[u_{\text{inc}}(k_0 r, \theta) + u_s(k_0 r, \theta) \right]_{r=a} = \frac{1}{\rho_p} \left[\frac{\partial u_{\text{int}}(k_p r, \theta)}{\partial r} \right]_{r=a} \tag{2.26}$$

which must now hold for the sum of the scattered and incident field, and the now existing internal field. ρ_0 and ρ_p characterize the density of the material outside and inside the sphere, and k_0 and k_p are the respective wave numbers. Please, note that we will always choose $\rho_0 = 1.0$ for the material outside the sphere! Note also that the size parameter inside the sphere with radius $r = a$ is denoted with

$$\beta_p = k_p \cdot a . \tag{2.27}$$

Since we are only interested in the scattered field we can first use condition (2.26) to express the coefficients g_{ln} of expansion (1.66) used for the internal field by the expansion coefficients c_{ln} and d_{ln} of the scattered and incident field. Inserting the resulting expression into condition (2.25) results again in relation (2.4) with the T-matrix for the sound penetrable sphere now given by

$$[T_p]_{nn'}^{ll'} = -\frac{j_n'(\beta_p) \cdot j_n(\beta) - \kappa \cdot j_n'(\beta) \cdot j_n(\beta_p)}{j_n'(\beta_p) \cdot h_n^{(1)}(\beta) - \kappa \cdot h_n'^{(1)}(\beta) \cdot j_n(\beta_p)} \cdot \delta_{ll'} \cdot \delta_{nn'} . \tag{2.28}$$

Here we have

$$\kappa = \frac{\kappa_r}{\kappa_k} ,$$ (2.29)

where

$$\kappa_r = \frac{\rho_p}{\rho_0} ,$$ (2.30)

and

$$\kappa_k = \frac{k_p}{k_0} .$$ (2.31)

Note that these ratios are used as input quantities in the Python program *centered_sphere*. The derivation, although a bit more elaborate, runs as before. That is, multiplication of each boundary condition with $\sin\theta \cdot Y^*_{l',n'}(\theta, \phi)$, subsequent integration from 0 to π with respect to θ and from 0 to 2π with respect to ϕ, and if taking the orthonormality relation of the spherical harmonics into account. In so doing, we have again to consider only the azimuthal mode $l = 0$, due to the incident field with expansion coefficients according to (1.70).

After juggling with some algebra, and if taking again relation (1.15) into account it follows that the corresponding elements of the S-matrix are given by

$$[S_p]_{nn} = \frac{[s]^{(1)}_{nn} + i \cdot [s]^{(2)}_{nn}}{[s]^{(1)}_{nn} - i \cdot [s]^{(2)}_{nn}} ,$$ (2.32)

where

$$[s]^{(1)}_{nn} = j'_n(\beta_p) \cdot y_n(\beta) - \kappa \cdot y'_n(\beta) \cdot j_n(\beta_p)$$ (2.33)

and

$$[s]^{(2)}_{nn} = j'_n(\beta_p) \cdot j_n(\beta) - \kappa \cdot j'_n(\beta) \cdot j_n(\beta_p)$$ (2.34)

are again two real-valued quantities if β and β_p are real-valued quantities! This can be compared to (2.15) so that the real-valued phase shift reads

$$\tan\Delta_n = \frac{[s]^{(2)}_{nn}}{[s]^{(1)}_{nn}} .$$ (2.35)

It follows that the S-matrix of the sound penetrable sphere is again a unitary matrix if β and β_p are real-valued quantities.

Results for the differential and total scattering cross-sections are shown in Figs. 2.7, 2.8 and 2.9. Especially in the backscattering region we can now observe an even more pronounced ripple structure for $\beta = 10, 20$. This prohibits once again the application of the GO approximation at higher size parameters. And, finally, Table 2.3 shows again that we are in agreement with the extinction paradox at very large size parameters.

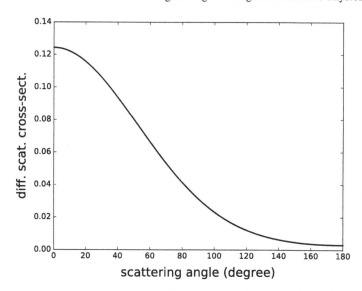

Fig. 2.7 Differential and total scattering cross-section of a sound penetrable sphere centered in the laboratory frame. Parameters: $\kappa_k = \rho_p = 1.5$, $a = 1.0\,\text{mm}$, $\beta = 1.0$, $ncut = 6$. Total scattering cross-section: $\sigma_{\text{tot}} = 0.5276$

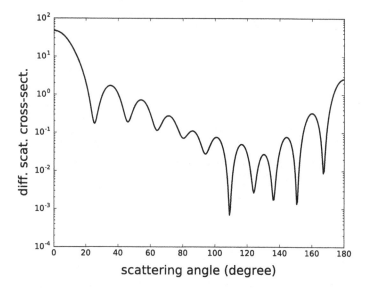

Fig. 2.8 Differential and total scattering cross-section of a sound penetrable sphere centered in the laboratory frame. Parameters: $\kappa_k = \rho_p = 1.5$, $a = 1.0\,\text{mm}$, $\beta = 10.0$, $ncut = 15$. Total scattering cross-section: $\sigma_{\text{tot}} = 8.632$

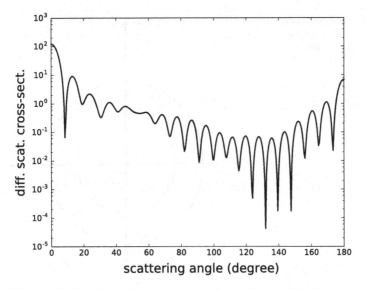

Fig. 2.9 Differential and total scattering cross-section of a sound penetrable sphere centered in the laboratory frame. Parameters: $\kappa_k = \rho_p = 1.5$, $a = 1.0\,\text{mm}$, $\beta = 20.0$, $ncut = 25$. Total scattering cross-section: $\sigma_{\text{tot}} = 6.860$

Table 2.3 Total scattering cross-section of a sound penetrable sphere with $\kappa_k = \rho_p = 1.5$ and radius $a = 1.0\,\text{mm}$. It approaches 2π for increasing size parameters	β	σ_{tot}
	1	0.5276
	50	6.766
	100	6.592 ($ncut = 115$)
	150	6.633 ($ncut = 165$)
	200	6.543 ($ncut = 215$)
	500	6.392 ($ncut = 515$)

2.2 T-Matrices of Two-Layered Spheres

Homogeneous sound soft, sound hard, and sound penetrable spheres are the most basic scattering geometries in spherical coordinates. A first step toward inhomogeneous scatterers are the concentric multilayered spheres centered in the laboratory frame. Here we will consider the two cases of spheres that consist of a sound soft or sound hard spherical core covered with a sound penetrable spherical layer (see Fig. 2.10). To determine the T-matrix expansions (1.69)/(1.70) and (1.67) are again used for the incident plane wave and the scattered field. But regarding the internal field in the sound penetrable layer, the radiating eigensolutions of the Helmholtz equation must now additionally be taken into account. That is, instead of (1.66) we have to use the expansion

Fig. 2.10 Inhomogeneous
two-layered sphere with a
sound soft or a sound hard
spherical core and a
penetrable layer of thickness
$ts = b - a$

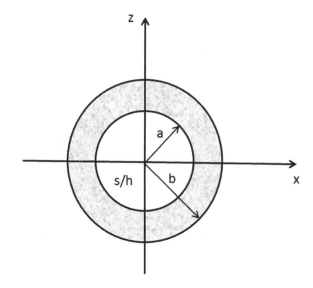

$$u_{\text{int}}(k_p r, \theta) = \sum_{n=0}^{ncut} \sum_{l=-n}^{n} g_{ln} \cdot \psi_n(k_p r, \theta) + f_{ln} \cdot \varphi_n(k_p r, \theta) \tag{2.36}$$

since the singular point $r = 0$ is excluded in this layer.

$$\beta_b = k_0 \cdot b \tag{2.37}$$

denotes the external size parameter of the two-layered sphere as a whole. And

$$\beta_p^{(a)} = k_p \cdot a \tag{2.38}$$

and

$$\beta_p^{(b)} = k_p \cdot b \tag{2.39}$$

are the two size parameters we have to consider at the two boundaries of the sound penetrable layer. Note also that we use the ratio (2.31) as the input quantity in the corresponding Python program *centered_2l_sphere.py*.

Depending on whether we consider the sound soft or sound hard core we first use condition (2.1) or (2.21) at the inner boundary $r = a$ to eliminate the coefficients f_{ln} in the expansion of the internal field. But, now, these conditions apply only to the internal field and not to the sum of the incident and scattered field! This gives

$$f_{ln} = \sum_{n'=0}^{ncut} \sum_{l'=-n'}^{n'} \left[T_{s/h}(\beta_p^{(a)}) \right]_{nn'}^{ll'} \cdot g_{l'n'} \cdot \tag{2.40}$$

$\left[T_{s/h}(\beta_p^{(a)})\right]_{nn'}^{ll'}$ therein are given by (2.5) and (2.22), respectively, but with β replaced by $\beta_p^{(a)}$ in the corresponding Bessel and Hankel functions. This relation is used next in boundary conditions (2.25) and (2.26) which must now hold at $r = b$. In close analogy to (2.28) we thus end up with the T-matrix

$$\left[T_{sp/hp}\right]_{nn'}^{ll'} = \frac{\left[\widetilde{j}_n(\beta_p^{(b)})\right]' \cdot j_n(\beta_b) - \kappa \cdot [j_n(\beta_b)]' \cdot \widetilde{j}_n(\beta_p^{(b)})}{\kappa \cdot \left[h_n^{(1)}(\beta_b)\right]' \cdot j_n(\beta_p^{(b)}) - \left[\widetilde{j}_n(\beta_p^{(b)})\right]' \cdot h_n^{(1)}(\beta_b)} \cdot \delta_{ll'} \cdot \delta_{nn'} \quad (2.41)$$

of the concentric two-layered sphere. $\widetilde{j}_n(\beta_p^{(a)})$ and $\left[\widetilde{j}_n(\beta_p^{(a)})\right]'$ therein are given by

$$\widetilde{j}_n(\beta_p^{(b)}) = j_n(\beta_p^{(b)}) + \left[T_{s/h}(\beta_p^{(a)})\right]_{nn} \cdot h_n^{(1)}(\beta_p^{(b)}) \quad (2.42)$$

and

$$\left[\widetilde{j}_n(\beta_p^{(b)})\right]' = j_n'(\beta_p^{(b)}) + \left[T_{s/h}(\beta_p^{(a)})\right]_{nn} \cdot \left[h_n^{(1)}(\beta_p^{(b)})\right]' . \quad (2.43)$$

Since this T-matrix exhibits the same structure as the T-matrix (2.28) of the sound penetrable sphere it is not too difficult to show that the corresponding S-matrix is again a unitary matrix. This proof is left to the reader as an exercise. The T-matrix of the sound soft or sound hard sphere results from (2.41) if $b = a$ is used. And the T-matrix of the sound penetrable sphere is approached if a tends to zero ($a = 0$ produces an error, due to the singular behavior of the Hankel function at this point).

Figures 2.11, 2.12 and 2.13 show the differential and total scattering cross-sections for the two-layered sphere with a sound soft core (sp-sphere) at the three size param-

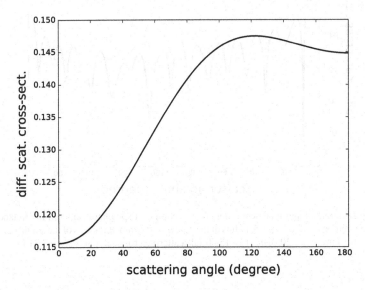

Fig. 2.11 Differential and total scattering cross-section of a sp-sphere centered in the laboratory frame. Parameters: $\kappa_k = \rho_p = 1.5$, radius of the core $a = 0.5$ mm, thickness of the sound penetrable layer $ts = 0.5$ mm, $\beta_b = 1.0$, $ncut = 15$. Total scattering cross-section: $\sigma_{\text{tot}} = 1.749$

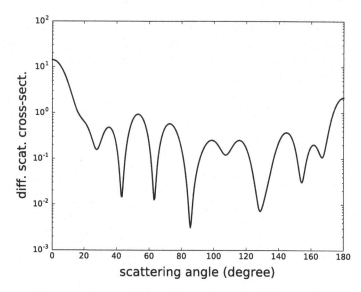

Fig. 2.12 Differential and total scattering cross-section of a sp-sphere centered in the laboratory frame. Parameters: $\kappa_k = \rho_p = 1.5$, radius of the core $a = 0.5$ mm, thickness of the sound penetrable layer $ts = 0.5$ mm, $\beta_b = 10.0$, $ncut = 15$. Total scattering cross-section: $\sigma_{\text{tot}} = 4.163$

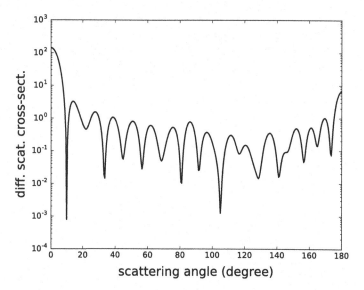

Fig. 2.13 Differential and total scattering cross-section of a sp-sphere centered in the laboratory frame. Parameters: $\kappa_k = \rho_p = 1.5$, radius of the core $a = 0.5$ mm, thickness of the sound penetrable layer $ts = 0.5$ mm, $\beta_b = 20.0$, $ncut = 15$. Total scattering cross-section: $\sigma_{\text{tot}} = 7.313$

Table 2.4 Total scattering cross-section of a sp-sphere with $\kappa_k = \rho_p = 1.5$, $a = 0.5\,\text{mm}$, and $ts = 0.5\,\text{mm}$

β_b	σ_{tot}
1	1.749
50	5.515
100	6.844 ($ncut = 115$)
150	6.910 ($ncut = 165$)
200	5.931 ($ncut = 215$)
500	6.080 ($ncut = 515$)

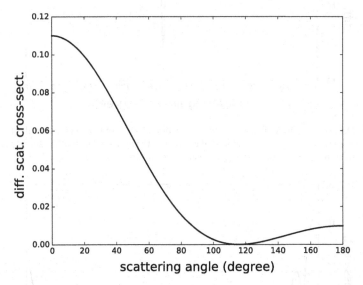

Fig. 2.14 Differential and total scattering cross-section of a hp-sphere centered in the laboratory frame. Parameters: $\kappa_k = \rho_p = 1.5$, radius of the core $a = 0.5\,\text{mm}$, thickness of the sound penetrable layer $ts = 0.5\,\text{mm}$, $\beta_b = 1.0$, $ncut = 15$. Total scattering cross-section: $\sigma_{tot} = 0.312$

eters β_b used before for the homogeneous spheres. Especially the behavior shown in Fig. 2.11 differs significantly from that of the corresponding sound soft and sound penetrable sphere (compare with Figs. 2.1 and 2.7). Regarding the higher size parameters we can again observe a pronounced ripple structure in the whole scattering plane. The total cross-sections of that sphere for increasing size parameters are given in Table 2.4. The corresponding results for the combination of a sound hard core and a sound penetrable layer (hp-sphere) are shown in Figs. 2.14, 2.15, 2.16 and Table 2.5. To approach the extinction paradox we now need much larger size parameters for both cases than we had for the three types of homogeneous spheres. All computations have been performed with the program *centered_2l_sphere.py*.

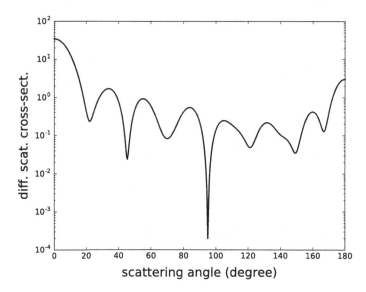

Fig. 2.15 Differential and total scattering cross-section of a hp-sphere centered in the laboratory frame. Parameters: $\kappa_k = \rho_p = 1.5$, radius of the core $a = 0.5\,\text{mm}$, thickness of the sound penetrable layer $ts = 0.5\,\text{mm}$, $\beta_b = 10.0$, $ncut = 15$. Total scattering cross-section: $\sigma_{\text{tot}} = 7.398$

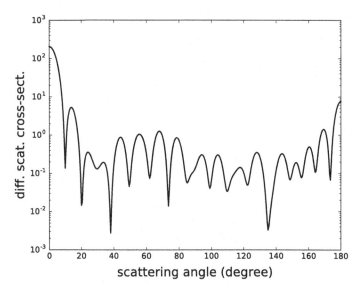

Fig. 2.16 Differential and total scattering cross-section of a hp-sphere centered in the laboratory frame. Parameters: $\kappa_k = \rho_p = 1.5$, radius of the core $a = 0.5\,\text{mm}$, thickness of the sound penetrable layer $ts = 0.5\,\text{mm}$, $\beta_b = 20.0$, $ncut = 15$. Total scattering cross-section: $\sigma_{\text{tot}} = 8.975$

Table 2.5 Total scattering cross-section of a hp-sphere with $\kappa_k = \rho_p = 1.5$, $a = 0.5\,\text{mm}$, and $ts = 0.5\,\text{mm}$

β_b	σ_{tot}
1	0.312
50	7.386
100	6.477 ($ncut = 115$)
150	6.130 ($ncut = 165$)
200	6.880 ($ncut = 215$)
500	6.688 ($ncut = 515$)

2.3 Rotated and Shifted Spheres

With the considerations of this section we want to convince ourselfs that a rotation or a local shift of the spherical scatterer within the laboratory frame does not change its scattering behavior in the far-field. This knowledge can be used with benefit to test the correct numerical implementation of the matrix of rotation and the separation matrix, and to estimate their range of applicability with respect to the size parameter and the distance of the shift. In so doing, we distinguish between a shift along the z-axis of the laboratory frame and an arbitrary shift. Besides the simplifications that arise for the former case, this distinction allows us to test the reciprocity condition for the bisphere configurations in the last chapter of this book by using two independent approaches.

2.3.1 Rotated Spheres

Regarding homogeneous or concentric multilayered spherical scatterers centered in the laboratory frame it is quite obvious that a rotation does not change the scattering behavior. It is therefore clear from the very beginning that we should get the same results. However, it provides us with the possibility to practice this transformation since we have to apply it frequently in the next chapters to more complex scattering configurations. Such a rotation is accomplished by the following three steps:

- The known expansion of the primary incident plane wave in the laboratory frame is transformed into the rotated system.
- Choosing a corresponding expansion of the scattered field in the rotated system, its unknown expansion coefficients are determined next by application of the respective boundary conditions.
- The expansion of the scattered field is finally transformed back into the laboratory frame, and the scattering quantities are calculated.

The first step was already considered in Sect. 1.3.1 (compare (1.71) and (1.72)) therein). Next we choose

$$u_s(k_0\bar{r}, \bar{\theta}, \bar{\phi}) = \sum_{n=0}^{ncut} \sum_{\bar{l}=-n}^{n} \widehat{c}_{\bar{l}n} \cdot \varphi_{\bar{l}n}(k_0\bar{r}, \bar{\theta}, \bar{\phi}) \tag{2.44}$$

as expansion of the scattered field in the rotated system. Applying the boundary conditions in a way that we have already discussed in the previous sections results in

$$\widehat{c}_{\bar{l}n} = [T]_{nn} \cdot \widehat{d}_{\bar{l}n} \tag{2.45}$$

for the expansion coefficients. $[T]_{nn}$ are just the diagonal elements of one of the T-matrices we derived for the different types of spheres centered in the laboratory frame. Note that the unnecessary summations and the corresponding Kronecker's in the "inflated" expressions of these T-matrices have now been omitted for reasons of readability. The final transformation of (2.44) back to the laboratory frame is performed by using (1.38). We thus get

$$u_s(k_0 r, \theta, \phi) = \sum_{n=0}^{ncut} \sum_{l=-n}^{n} c_{ln} \cdot \varphi_{ln}(k_0 r, \theta, \phi) \tag{2.46}$$

for the scattered field, and with expansion coefficients now given by

$$c_{ln} = [T_L]_{nn}^{l0} \cdot d_n . \tag{2.47}$$

$$[T_L]_{nn}^{l0} = \sum_{\bar{l}=-n}^{n} D_{l\bar{l}}^{(n)}(\alpha, \theta_p, 0) \cdot [T]_{nn} \cdot D_{\bar{l}0}^{(n)}(0, -\theta_p, -\alpha) \tag{2.48}$$

are the elements of a new T-matrix in the laboratory frame. That is, the initial transformation into the rotated system and the final transformation back to the laboratory frame is condensed into this new T-matrix $\mathbf{T_L}$. Taking the far-field limit of the radiating eigensolutions $\varphi_{ln}(k_0 r, \theta, \phi)$ into account we are then able to calculate the differential and total scattering cross-sections as described in Sect. 1.3.1.

This transformation is implemented in program *rotated_sphere.py* for sound soft, hard, and penetrable spheres. Running it with the parameters used in Figs. 2.1, 2.2, 2.3, 2.4, 2.5, 2.6, 2.7, 2.8 and 2.9 produces the same results, and independent of the choice of the Eulerian angles α and θ_p of rotation. However, this program is not really needed for this proof. Having in mind the unitarity property of the matrix of rotation it follows that

$$\sum_{\bar{l}=-n}^{n} D_{l\bar{l}}^{(n)}(\alpha, \theta_p, 0) \cdot D_{\bar{l}0}^{(n)}(0, -\theta_p, -\alpha) = \delta_{l0} \tag{2.49}$$

must hold for the summation with respect to the product of the matrices of rotation in (2.48). And since the diagonal elements $[T]_{nn}$ are not dependent on the summation index $\bar{l}$ it follows in a straightforward way that the new T-matrix $\mathbf{T_L}$ will be identical with the T-matrices of the spheres we derived in the previous sections. The program *rotated_sphere* is therefore essentially a numerical test of the unitarity property of the matrix of rotation. It should also be noted that expressions like (2.46) can identically be rewritten as follows:

$$u_s(k_0 r, \theta, \phi) = \sum_{l=-ncut}^{ncut} \sum_{n=|l|}^{ncut} c_{ln} \cdot \varphi_{ln}(k_0 r, \theta, \phi) . \qquad (2.50)$$

This reordering of the summation is frequently applied in the text as well as in the Python programs. It will allow us moreover to introduce a second truncation parameter $lcut \leq ncut$. We will benefit from this possibility in the next chapter when dealing with arbitrarily rotated Janus spheres.

2.3.2 Spheres Shifted Along the z-Axis

To describe the scattering behavior of a sphere shifted along the positive z-axis of the laboratory frame we have to accomplish the following three steps:

- The known expansion of the primary incident plane wave in the laboratory frame is transformed into the shifted system.
- Choosing a corresponding expansion of the scattered field, its unknown expansion coefficients are determined next by application of the boundary conditions in the shifted system.
- The expansion of the scattered field is finally transformed back into the laboratory frame, and the scattering quantities are calculated.

The first step can again be found in Sect. 1.3.1 (compare (1.73) and (1.74) therein). Next we use

$$u_s(k_0 r_1, \theta_1, \phi_1) = \sum_{n'=0}^{n'cut} \widehat{c}_{n'} \cdot \varphi_{n'}(k_0 r_1, \theta_1, \phi_1) \qquad (2.51)$$

for the scattered field in the shifted frame. Applying the boundary conditions in the shifted frame results now in

$$\widehat{c}_{n'} = [T]_{n'n'} \cdot \widehat{d}_{n'} \qquad (2.52)$$

for the expansion coefficients, where $[T]_{n'n'}$ are again identical with the diagonal elements of the T-matrix we derived already for a certain type of spheres centered in the laboratory frame. The transformation of (2.51) with coefficients (2.52) back into the laboratory frame is accomplished by using (1.28) but with interchanged arguments (r, θ, ϕ) and (r_1, θ_1, ϕ_1), and by taking (1.30) into account. Moreover,

(1.28) is used since we are only interested in the far-field of the laboratory frame where we have $r > b$. This results in

$$u_s(k_0 r, \theta, \phi) = \sum_{\nu=0}^{\nu cut} c_\nu \cdot \varphi_\nu(k_0 r, \theta, \phi) \tag{2.53}$$

with coefficients c_ν now given by

$$c_\nu = \sum_{n'=0}^{n'cut} \sum_{n=0}^{ncut} (-1)^\nu \cdot \widehat{S}^0_{\nu n'}(k_0 b) \cdot [T]_{n' n} \cdot (-1)^n \cdot \widehat{S}^0_{n n'}(k_0 b) \cdot d_n . \tag{2.54}$$

Application of the sum rule (1.80) results in a further simplification of this expression:

$$c_\nu = e^{i k_0 b} \cdot \sum_{n'=0}^{n'cut} (-1)^\nu \cdot \widehat{S}^0_{\nu n'}(k_0 b) \cdot [T]_{n' n'} \cdot d_{n'} . \tag{2.55}$$

Introducing the new T-matrix

$$[T_L]_{\nu n'} = e^{i k_0 b} \cdot (-1)^\nu \cdot \widehat{S}^0_{\nu n'}(k_0 b) \cdot [T]_{n' n'} \tag{2.56}$$

(2.55) can be rewritten into

$$c_\nu = \sum_{n'=0}^{n'cut} [T_L]_{\nu n'} \cdot d_{n'} . \tag{2.57}$$

The transformation into the shifted system and the transformation back into the laboratory frame is again condensed into this new T-matrix. Note also that only the azimuthal mode $l = 0$ is needed throughout this transformation process!

We are now in the position to verify numerically if a local shift of a sphere along the z-axis of the laboratory frame does not change its scattering behavior in the far-field. This can be accomplished with the Python program *z_shifted_sphere*. It can be seen from (2.53) and (2.55) that we have now to consider the two truncation parameters νcut and $n'cut$, in general. But in all our Python programs with multiple truncation parameters we will use only one truncation parameter. I.e., $\nu cut = n'cut = ncut$ was used in program *z_shifted_sphere*. The differential and total scattering cross-sections for a sound soft sphere centered in the laboratory frame and if shifted along the positive z-axis is depicted in Fig. 2.17. The results for the centered sphere were obtained with the program *centered_sphere* while the program *z_shifted_sphere* was used for the shifted sphere. Regarding the shifted sphere, the computations were performed at three different values of the truncation parameter $ncut$ but with the shift kept fix. We can see that the results of the centered sphere are indeed reproduced by the shifted sphere but only if using a much larger truncation parameter (see the lower

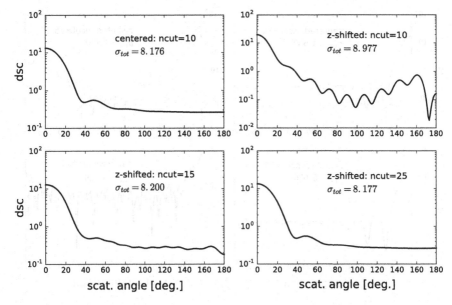

Fig. 2.17 Differential and total scattering cross-sections of a sound soft sphere. Parameters: radius $a = 1.0$ mm, size parameter of $\beta = 5.0$. Calculations have been performed for the centered sphere and for the sphere shifted by $b = 3.0$ mm along the positive z-axis—for the latter at 3 different truncation parameters $ncut$

right subplot). It is a further consequence of this larger truncation parameter and the more complex transformation procedure that the latter computations require much more computing time.

Another example for a sound hard sphere at a size parameter of $\beta = 10$ and for the same shift that was used in Fig. 2.17 is presented in Fig. 2.18. Now we are not able to reproduce the differential scattering cross-section of the centered sphere in the whole scattering plane. Increasing the truncation parameter beyond the value of $ncut = 37$ produce totally wrong results, caused by the inaccuracy of the separation matrix. A much greater programming effort would be required to make the separation matrix running within a sufficient accuracy at such larger truncation parameters. However, from the lower left subplot we can see that we are at least able to reproduce the total as well as the differential scattering cross-section within a sufficient accuracy in the near forward direction up to a scattering angle of $\theta = 60°$. This example was chosen to demonstrate that the derived mathematical expressions suggest a flexibility with respect to the parameters that is rarely guaranteed in practical computations. To identify the range of applicability of a certain method and its numerical implementation it is therefore of some importance to have access to known and accurate solutions and to perform such intercomparisons. This is moreover the only way to gain trust into a software that was overtaken from a third party without knowing its methical and numerical details.

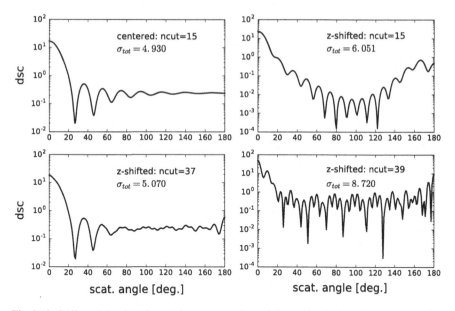

Fig. 2.18 Differential and total scattering cross-sections of a sound hard sphere. Parameters: radius $a = 1.0$ mm, size parameter of $\beta = 10.0$. Calculations have been performed for the centered sphere and for the sphere shifted by $b = 3.0$ mm along the positive z-axis—for the latter at 3 different truncation parameters *ncut*

2.3.3 Arbitrarily Shifted Spheres

Let us now consider the more complex situation of an arbitrarily shifted sphere. As already discussed in Sect. 1.2.3, there are different ways to describe such a shift. Here we will put the focus on the first combination that consists of a rotation and a shift along the new z-axis after this rotation (see Fig. 1.11). An application of the second combination, that is in fact a translation of the laboratory frame, will add nothing new. This results from to the unitarity property (2.49) that applies also to the two additional rotations required in this second combination, and as long as a single sphere is considered. Only in the last chapter of this book, when we discuss arbitrarily located bispheres, we can really benefit from this second combination.

The first step consists again in the transformation of expansion (1.69)/(1.70) of the incident plane wave into the rotated and shifted system. This is accomplished by (1.47) and results in

$$u_{\text{inc}}(k_0 r'', \theta'', \phi'') = \sum_{n=0}^{ncut} \sum_{l'=-n}^{n} \sum_{n'=0}^{n'cut} \left[\widehat{T}_{\text{rs}}\right]_{nn'}^{0\,l'} \cdot d_n \cdot \psi_{l'\,n'}(k_0 r'', \theta'', \phi'') , \qquad (2.58)$$

where $\left[\widehat{T}_{rs}\right]_{n n'}^{0 \, l'}$ is given by (1.45). Using the reordering (2.50) we can reformulate this expression as follows:

$$u_{\text{inc}}(k_0 r'', \theta'', \phi'') = \sum_{l'=-ncut}^{ncut} \sum_{n'=0}^{n'cut} \widehat{d}_{l' \, n'} \cdot \psi_{l' \, n'}(k_0 r'', \theta'', \phi'') \qquad (2.59)$$

with the new expansion coefficients $\widehat{d}_{l' \, n'}$ given by

$$\widehat{d}_{l' \, n'} = \sum_{n=|l'|}^{ncut} \left[\widehat{T}_{rs}\right]_{n \, n'}^{0 \, l'} \cdot d_n \, . \qquad (2.60)$$

Regarding the scattered field in this local system we choose the expansion

$$u_{\text{s}}(k_0 r'', \theta'', \phi'') = \sum_{l'=-ncut}^{ncut} \sum_{n'=0}^{n'cut} \widehat{c}_{l' \, n'} \cdot \varphi_{l' \, n'}(k_0 r'', \theta'', \phi'') \qquad (2.61)$$

in terms of the radiating eigensolutions. In the next step we have to apply the boundary conditions for the respective sphere in this local system to determine the unknown expansion coefficients $\widehat{c}_{l' \, n'}$. Since the orthonormality relation (1.14) for the spherical harmonics holds also in this system we get

$$\widehat{c}_{l' \, n'} = [T]_{n' \, n'} \cdot \widehat{d}_{l' \, n'} \qquad (2.62)$$

with $[T]_{n' n'}$ representing again one of the T-matrices derived in the first two sections of this chapter for the respective spheres centered in the laboratory frame. And, finally, using (1.52) leads to the following expression for the scattered field in the laboratory frame:

$$u_{\text{s}}(k_0 r, \theta, \phi) = \sum_{n''=0}^{n''cut} \sum_{l''=-n''}^{n''} c_{l'' n''} \cdot \varphi_{l'' n''}(k_0 r, \theta, \phi) \, ; \ r > b \, , \qquad (2.63)$$

where

$$c_{l'' n''} = \sum_{l'=-ncut}^{ncut} \sum_{n'=0}^{n'cut} \widehat{c}_{l' \, n'} \cdot \left[\widehat{T}_{rs}^{-1}\right]_{n' \, n''}^{l' \, l''} \, . \qquad (2.64)$$

Introducing the new T-matrix

$$[T_L]_{n'' n'}^{l'' l'} = [T]_{n' \, n'} \cdot \left[\widehat{T}_{rs}^{-1}\right]_{n' \, n''}^{l' \, l''} \, . \qquad (2.65)$$

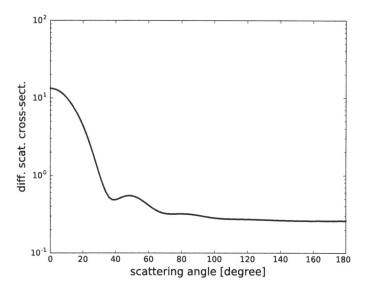

Fig. 2.19 Differential and total scattering cross-sections of a sound soft sphere. Parameters: radius $a = 1.0$ mm, size parameter $\beta = 5.0$, Eulerian angles of rotation $\alpha = \theta_p = 45°$, shift along the new z-axis after rotation $b = 3.0$ mm, truncation parameter $ncut = 25$. Total scattering cross-section: $\sigma_{\text{tot}} = 8.174$

in the laboratory frame the expansion coefficients (2.64) can be rewritten into

$$c_{l'' n''} = \sum_{l'=-ncut}^{ncut} \sum_{n'=0}^{n'cut} [T_L]_{n'' n'}^{l'' l'} \cdot \widehat{d}_{l' n'} . \tag{2.66}$$

This much more complex expression should produce the same differential and total scattering cross-sections as the simple expression for the respective sphere that is centered in the laboratory frame. That this is indeed the case can be seen from Fig. 2.19 for the sound soft sphere of Fig. 2.17. The computation was performed with the Python program *shifted_sphere*. Note that only one truncation parameter $n''cut = n'cut = ncut$ is used in this program.

2.4 Boundary Conditions for the Debye Potentials and an Outlook to Scattering from Nonspherical Geometries

It was already discussed at the end of the previous chapter that the electromagnetic wave scattering problem on spherical objects can be reduced to the solution of two scalar problems by using the Debye potentials. We now want to continue this discussion with introducing the corresponding boundary conditions for the two basic scattering objects in this context—the ideal metallic and the homogeneous dielectric sphere. The boundary conditions for the Debye potentials are a direct consequence of the boundary conditions required for the tangential components (the θ- and ϕ-components in the case of spherical scatterers) of the electromagnetic field on the spherical surface of the respective scatterer. For the ideal metallic sphere with radius $r = a$ the condition for the tangential components of the total electric field $\vec{E}_t = \vec{E}_{\text{inc}} + \vec{E}_s$ (no internal field exists in this case) reads

$$\hat{r} \times \vec{E}_t(a, \theta, \phi) = 0 \tag{2.67}$$

while

$$\frac{\partial(r \cdot \Pi_e^s(r, \theta, \phi))}{\partial r} + \frac{\partial(r \cdot \Pi_e^{\text{inc}}(r, \theta, \phi))}{\partial r} = 0 \; ; \; r = a \tag{2.68}$$

$$\Pi_m^s(a, \theta, \phi) + \Pi_m^{\text{inc}}(a, \theta, \phi) = 0 \tag{2.69}$$

hold for the corresponding Debye potentials. Regarding the dielectric sphere with radius $r = a$ we have on the other hand

$$\hat{r} \times \left[\vec{E}_t(a, \theta, \phi) - \vec{E}_{\text{int}}(a, \theta, \phi) \right] 0 = 0 \tag{2.70}$$

$$\hat{r} \times \left[\vec{H}_t(a, \theta, \phi) - \vec{H}_{\text{int}}(a, \theta, \phi) \right] 0 = 0 \tag{2.71}$$

for the tangential components of the total external and internal electromagnetic fields, and

$$\frac{\partial(r \cdot \Pi_m^s(r, \theta, \phi))}{\partial r} + \frac{\partial(r \cdot \Pi_m^{\text{inc}}(r, \theta, \phi))}{\partial r} =$$
$$\frac{\partial(r \cdot \Pi_m^{\text{int}}(r, \theta, \phi))}{\partial r} \; ; \; r = a \tag{2.72}$$

$$\frac{\partial(r \cdot \Pi_e^s(r, \theta, \phi))}{\partial r} + \frac{\partial(r \cdot \Pi_e^{\text{inc}}(r, \theta, \phi))}{\partial r} =$$
$$\frac{\epsilon_0}{\epsilon} \cdot \frac{\partial(r \cdot \Pi_e^{\text{int}}(r, \theta, \phi))}{\partial r} \; ; \; r = a \tag{2.73}$$

$$\Pi_m^s(a, \theta, \phi) + \Pi_m^{\text{inc}}(a, \theta, \phi) = \Pi_m^{\text{int}}(a, \theta, \phi) \tag{2.74}$$

$$\Pi_e^s(a, \theta, \phi) + \Pi_e^{\text{inc}}(a, \theta, \phi) = \Pi_e^{\text{int}}(a, \theta, \phi) \tag{2.75}$$

for the Debye potentials. ϵ_0 and ϵ are the dielectric constants outside and inside the sphere. Using expansions (1.108)–(1.113) in the boundary conditions for the Debye potentials, and if applying the same procedure that was used to determine the T-matrices of the sound soft, sound hard and sound penetrable sphere we end up with the following expansion coefficients a_n and b_n for the Debye potentials Π_e^s and Π_m^s related to the scattered electromagnetic field:

- ideal metallic sphere:

$$a_n = -i^n \frac{2n+1}{n(n+1)} \cdot \frac{\frac{\partial}{\partial r}[r \cdot j_n(k_0 r)]_{r=a}}{\frac{\partial}{\partial r}\left[r \cdot h_n^{(1)}(k_0 r)\right]_{r=a}} \tag{2.76}$$

$$b_n = -i^n \frac{2n+1}{n(n+1)} \cdot \frac{j_n(k_0 a)}{h_n^{(1)}(k_0 a)} \; . \tag{2.77}$$

- homogeneous dielectric sphere:

$$a_n = i^n \frac{2n+1}{n(n+1)} \cdot \frac{\epsilon_0 j_n^{(0)}\left[j_n^{(s)}\right]' - \epsilon j_n^{(s)}\left[j_n^{(0)}\right]'}{\epsilon j_n^{(s)}[h_n]' - \epsilon_0 h_n\left[j_n^{(s)}\right]'} \tag{2.78}$$

$$b_n = i^n \frac{2n+1}{n(n+1)} \cdot \frac{j_n^{(0)}\left[j_n^{(s)}\right]' - j_n^{(s)}\left[j_n^{(0)}\right]'}{j_n^{(s)}[h_n]' - h_n\left[j_n^{(s)}\right]'} \tag{2.79}$$

Note that the following quantities are used in (2.78) and (2.79):

$$j_n^{(0)} = j_n(k_0 a) \tag{2.80}$$

$$j_n^{(s)} = j_n(k a) \tag{2.81}$$

$$h_n = h_n^{(1)}(k_0 a) \tag{2.82}$$

$$\left[j_n^{(0)}\right]' = \frac{\partial}{\partial r}[r \cdot j_n(k_0 r)]_{r=a} \tag{2.83}$$

$$\left[j_n^{(s)}\right]' = \frac{\partial}{\partial r}[r \cdot j_n(k r)]_{r=a} \tag{2.84}$$

$$[h_n]' = \frac{\partial}{\partial r}\left[r \cdot h_n^{(1)}(k_0 r)\right]_{r=a} \; . \tag{2.85}$$

These coefficients and the resulting scattering quantities are well documented in the literature (see Chap. 9.22 in [6], for example). There are several freely available Python programs to calculate the corresponding scattering quantities from these coefficients (see the code *miepython* one can find in [7], for example). But let us now discuss a total different aspect:

The T-matrix method for nonspherical scattering objects was originally introduced by Waterman and applied to acoustic scattering problems in [8]. He was using the so-called "Extended Boundary Condition" as the key condition to circumvent the singularity problem at the inner eigenresonances of the scatterer that was discussed (sometimes quite controversely) in other methods. But there exists a simpler way to end up with the same T-matrices, as could be demonstrated later (see [2] for more details). At the end of this section we will therefore briefly sketch the application of the procedure described in the first section of this chapter to derive the T-matrices of nonspherical but rotational symmetric geometries. Two parameter representations of such surfaces have already been introduced in (1.5) and (1.6).

Regarding a rotational symmetric sound soft scatterer that is axisymmetrically oriented in the laboratory frame (i.e., the z-axis of the laboratory frame corresponds with its axis of rotation) the boundary condition (2.1) now becomes

$$u_{\text{inc}}(k_0 r(\theta), \theta) + u_s(k_0 r(\theta), \theta) = 0 \tag{2.86}$$

with $r(\theta)$ representing the boundary surface. Using series expansions (1.67) and (1.69)/(1.70) we can thus write

$$\sum_{n=0}^{ncut} d_n \cdot \psi_n(k_0 r(\theta), \theta) + \sum_{n=0}^{ncut} c_n \cdot \varphi_n(k_0 r(\theta), \theta) = 0 . \tag{2.87}$$

Note that $l = 0$ is again the only azimuthal mode that must be considered! In the next step we multiply this equation with at first arbitrary weighting functions $w_{n'}(k_0 r(\theta), \theta, \phi)$ with n' running from $0, \cdots ncut$ and integrate with respect to dS (with dS calculated according to (1.7)!) afterwards. This results in the following matrix equation for the unknown expansion coefficients c_n of the scattered field:

$$\mathbf{B} \cdot \vec{c} = -\mathbf{A} \cdot \vec{d} . \tag{2.88}$$

$\mathbf{A}$ and $\mathbf{B}$ are $(ncut + 1) \times (ncut + 1)$ square matrices with elements

$$[A]_{n' n} = \oint_S w_{n'}(k_0 r(\theta), \theta, \phi) \cdot \psi_n(k_0 r(\theta), \theta, \phi) \, dS \tag{2.89}$$

$$[B]_{n' n} = \oint_S w_{n'}(k_0 r(\theta), \theta, \phi) \cdot \varphi_n(k_0 r(\theta), \theta, \phi) \, dS , \tag{2.90}$$

and $\vec{c}$ and $\vec{d}$ are vectors with the expansion coefficients of the scattered and incident field as components. We thus have finally

$$\vec{c} = -\mathbf{B}^{-1} \cdot \mathbf{A} \cdot \vec{d} = \mathbf{T_S} \cdot \vec{d} . \tag{2.91}$$

$\mathbf{T_S} = -\mathbf{B}^{-1} \cdot \mathbf{A}$ represents the T-matrix of the rotational symmetric sound soft scatterer in axisymmetric orientation. The precise expression of this T-matrix is dependent on the choice of the weighting functions used in the surface integrals (2.89) and (2.90), of course. These integrals must be calculated numerically, in general. The choice of the weighting functions has a major impact on the convergence behavior and the numerical stability of the solution since we have now to invert a full matrix—the matrix $\mathbf{B}$. As a consequence, the diagonal character of the T-matrix gets lost, and the expansion coefficients of the scattered field are no longer final, in general. However, if choosing the eigensolutions $\psi_n(k_0 r(\theta), \theta, \phi)$ or $\varphi_n(k_0 r(\theta), \theta, \phi)$ as weighting functions, and if the spherical surface is considered we end up again with the T-matrix (2.5). The same way can be used to derive the T-matrices of sound hard and sound penetrable but rotational symmetric scatterers in axisymmetric orientation. But what happens if the scatterer is not in an axisymmetric orientation with respect to the laboratory frame? This problem can again be solved by applying a rotation of the laboratory frame, as described in the first chapter of this book. This rotation must be performed in such a way that the new z'-axis corresponds again with the axis of symmetry of the scatterer. Then we can proceed in the same way as described above. That we have now to take the azimuthal modes of this rotation into account is the only difference. The restriction to rotational symmetric surfaces has the advantage that it results in a T-matrix that exhibits a block-diagonal structure with respect to these azimuthal modes. It should also be emphasized that (2.88) can be used as an appropriate starting point for an iterative solution if the boundary surface of the scatterer deviates only slightly from that of a sphere. In so doing, one can avoid the inversion of a full matrix, as demonstrated in [9]. A more detailed discussion of this approach and its relation to other solution methods developed for nonspherical scattering problems can be found in [2].

2.5   Python Programs

All the Python programs used in this chapter to calculate the differential and total scattering cross-sections ar organized as follows: The main program, denoted with {*name*}.*py*, is separated from the program {*name*}_*input.py* which must be used first to generate the input file *input_data_{name}.txt*. The differential scattering cross-sections can be found in the file *dscross_{name}.txt*. Just an example to avoid confusion: To calculate the differential and total scattering cross-section of a sound soft sphere centered in the laboratory frame use *centered_sphere_input.py* first to generate the file *input_data_centered_sphere.txt* that contains all input parameters. Then start *centered_sphere.py* to perform the computation. The differential scattering cross-sections are finally written to the file *dscross_centered_sphere.txt*. Please, have also in mind that the modul *basics.py*, that was described in the previous chapter, must exist in the actual working directory.

Regarding the aspect of choosing the truncation parameter *ncut* appropriately, the following criterion has been found useful in [2, 10, 11], for example: Calculate the differential scattering cross-section for increasing values of *ncut* and in definite steps in the whole scattering plane $\theta \in [0, \pi]$. If the relative error of two successive computations is less than a certain threshold in 80% of the scattering angles the last value of *ncut* is taken as the convergence parameter. The restriction to 80% of the scattering angles avoids exaggerate accuracy requirements in the deep down spikes of the differential scattering cross-sections.

The Python programs of this chapter are the following:

- programs *centered_sphere.py* and *centered_sphere_input.py*:
  These programs can be used to calculate the differential and total scattering cross-sections of sound soft, sound hard, and sound penetrable spheres centered in the laboratory frame at a given size parameter β. Note that the truncation parameter

$ncut$ is fixed to $ncut = \beta + 5$ in *centered_sphere.py*. This produces exact results for size parameters not exceeding $\beta = 20$. This choice of $ncut$ must possibly be increased for larger size parameters.

- programs *centered_2l_sphere.py* and *centered_2l_sphere_input.py*:
  These programs can be used to calculate the differential and total scattering cross-sections of a two-layered sphere with a concentric sound soft or sound hard core and a sound penetrable shell centered in the laboratory frame. The computation is performed at a given size parameter β. The truncation parameter $ncut$ must be given by the user. It depends strongly not only on the size parameter but also on the thickness of the sound penetrable shell and the material parameters.
- programs *rotated_sphere.py* and *rotated_sphere_input.py*:
- programs *z_shifted_sphere.py* and *z_shifted_sphere_input.py*:
- programs *shifted_sphere.py* and *shifted_sphere_input.py*:
  These last three packages can be used for a numerical test of the fact that a rotated, a z-shifted, or an arbitrarily shifted sphere produces the same scattering behavior as the corresponding sphere that is centered in the laboratory frame. The application of these programs are restricted to sound soft, hard, and penetrable spheres. But it is straightforward to extend these programs to become applicable also to the two-layered spheres by adding the corresponding T-matrices which are included in the module *basics.py*. However, this task is left to the reader as an exercise.

Several examples of how to use this software have been already presented in this chapter. The full programs are given in Appendix B.

References

1. Colton, D., Kress, R.: Integral Equation Methods in Scattering Theory. Wiley, New York (1983)
2. Rother, T., Kahnert, M.: Electromagnetic Wave Scattering on Nonspherical Particles: Basic Methodology and Applications. Springer, Heidelberg (2014)
3. Nussenzveig, H.M.: High-frequency scattering by an impenetrable sphere. Ann. Phys. **34**, 23–955 (1965)
4. Turley, S.: Acoustic Scattering from a Sphere, found at: http://volta.byu.edu/winzip/scalar_sphere.pdf (2006)
5. Rother, T.: Green's Functions in Classical Physics. Springer International Publishing AG, Cham, Switzerland (2017)
6. van de Hulst, H.C.: Light Scattering by Small Particles. Dover, New York (1981)
7. https://github.com/scottprahl/miepython
8. Waterman, P.C.: New formulation of acoustic scattering. J. Acoust. Soc. Am. **45**, 1417–1429 (1969)
9. Rother, T., Wauer, J.: Case study about the accuracy behaviour of three different T-matrix methods. Appl. Opt. **49**, 5746–5756 (2010)
10. Barber, P.W., Hill, S.C.: Light Scattering by Particles: Computational Methods. World Scientific, Singapore (1990)
11. Wiscombe, J.A., Mugnai, A.: Single scattering from nonspherical Chebyshev particles. NASA Reference Publ., vol. 1157 (1986)

Chapter 3
Scattering on Janus Spheres

Janus particles—named after the two-faced god of the ancient Roman religion—are objects with interesting properties. They are of growing importance in many fields of technology and hold an enormous potential of new applications in material sciences, biotechnology, and in nanotechnology, for example (for an overview, see [5]). Janus particles are already industrially manufactured. Their scattering properties, among others, are of some importance not only for diagnostic purposes during the production process but also for discovering new applications. It will therefore be demonstrated in this chapter that a slight modification of the T-matrix method, that was introduced in the previous chapter, can successfully be applied to study the scattering behavior of Janus particles. Although again focusing on spherical Janus particles it should be mentioned that this modified T-matrix method can be generalized to become applicable also to nonspherical geometries, according to the remarks in Sect. 2.4.

Figure 3.1 represents the geometry of the Janus sphere we have in mind. It consists of a homogeneous spherical basis of radius $r = a$ whose surface is split into two different regions. This splitting is controlled by the splitting angle θ_J. That is, for the sake of simplicity we will consider only rotational symmetric surface regions. It is further assumed that different boundary conditions apply to the two surface regions. In view of the boundary conditions (2.1), (2.21) and (2.25)/(2.26) there are three different types of Janus spheres. The combinations sound hard/sound soft, sound soft/sound penetrable, and sound hard/sound penetrable are the relevant boundary conditions for the Janus sphere of h-s, s-p, and h-p type with the first letter denoting the boundary condition that holds on the upper surface region marked by the dashed line in Fig. 3.1. These three types of Janus spheres will be considered in detail in what follows. However, searching the web for Janus spheres one can find another

Electronic supplementary material The online version of this chapter (https://doi.org/10.1007/978-3-030-36448-9_3) contains supplementary material, which is available to authorized users.

Fig. 3.1 Janus sphere

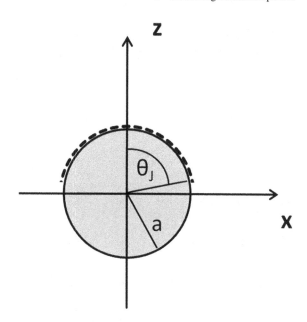

understanding of this object. For example, two homogeneous hemispheres that are made of different materials and stuck together are also called Janus spheres. But such objects are outside the scope of this book. It should also be mentioned that one can find only very few information in the web regarding the scattering behavior of such Janus spheres, so far. Therefore, with the considerations of this chapter we enter uncharted waters. The presented results are aimed at improving this situation and providing scattering data for intercomparison purposes with experiments and with results from other solution methods.

3.1 Janus Sphere of h-s Type

3.1.1 Axisymmetric Orientation

We start our discussion with the most simple Janus sphere—the Janus sphere of h-s type in axisymmetric orientation. That is, it is again assumed that the incident plane wave travels along the positive z-direction of the laboratory frame, and that this axis corresponds with the axis of rotational symmetry of the Janus sphere. In so doing, we can again benefit from the fact that only the azimuthal mode $l = 0$ must be taken into account in all the corresponding series expansions of the fields.

All the Janus spheres of our interest exhibit a sharp transition between the two different surface regions. But this sharp splitting can be considered as a limiting result of a smooth and differentiable generalized Robin boundary condition

$$f_<(\theta) \cdot u_<(a, \theta) + f_>(\theta) \cdot u_>(a, \theta) = 0 \tag{3.1}$$

that is assumed to hold on the whole surface at $r = a$ of the respective Janus sphere. The two functions $f_<(\theta)$ and $f_>(\theta)$ therein are the Fermi functions

$$f_<(\theta) = \left[e^{\frac{(\theta - \theta_J)}{\epsilon}} + 1 \right]^{-1} \tag{3.2}$$

and

$$f_>(\theta) = \left[e^{\frac{(\theta_J - \theta)}{\epsilon}} + 1 \right]^{-1} . \tag{3.3}$$

These Fermi functions tend to the Heaviside step function in the limiting case $\epsilon \to 0$, i.e., we have

$$\lim_{\epsilon \to 0} f_<(\theta) = H(\theta_J - \theta) \tag{3.4}$$

and

$$\lim_{\epsilon \to 0} f_>(\theta) = H(\theta - \theta_J) . \tag{3.5}$$

This results in the sharp surface splitting together with the required boundary conditions on each subsurface. Let us apply this idea to the Janus sphere of h-s-type. That is, the boundary condition of the sound hard sphere must hold on the dashed subregion in Fig. 3.1 while the boundary condition of the sound soft sphere applies to the other subregion.

Since there is no internal field that has to be considered the quantities $u_<$ and $u_>$ in (3.1) are given by

$$u_<(a, \theta) = \left[\frac{\partial u_t(r, \theta)}{\partial r} \right]_{r=a} \tag{3.6}$$

and

$$u_>(a, \theta) = u_t(a, \theta) \tag{3.7}$$

with u_t representing the sum of the incident plane wave and the scattered field. In this case, condition (3.1) looks like a generalized Robin boundary condition that combines the total outer field with its normal derivative on the spherical surface. Using expansions (1.67) and (1.69) together with the boundary conditions (2.1) and (2.21) in (3.1) provides

$$\sum_{n=0}^{ncut} c_n \cdot \left\{ \left[h_n^{(1)}(\beta) \right]' \cdot f_<(\theta) + h_n^{(1)}(\beta) \cdot f_>(\theta) \right\} \cdot Y_{0n}(\theta, \phi) =$$

$$- \sum_{n=0}^{ncut} d_n \cdot \left\{ j_n'(\beta) \cdot f_<(\theta) + j_n(\beta) \cdot f_>(\theta) \right\} \cdot Y_{0n}(\theta, \phi) . \tag{3.8}$$

This equation is multiplied with the spherical harmonics $Y_{0m}(\theta, \phi)$; $m = 0, \ldots,$ $ncut$ and integrated according to

$$\int_0^\pi \int_0^{2\pi} \sin\theta \ldots d\theta \, d\phi \tag{3.9}$$

afterwards. The following integrals remain for every finite θ_J with $0 < \theta_J < \pi$, and if taking the limit $\epsilon \to 0$ into account:

$$[Y_<]_{mn} = \frac{1}{2} \cdot \sqrt{(2m+1) \cdot (2n+1)} \cdot \int_0^{\theta_J} P_m(\cos\theta) \cdot P_n(\cos\theta) \cdot \sin\theta \, d\theta \tag{3.10}$$

and

$$[Y_>]_{mn} = \frac{1}{2} \cdot \sqrt{(2m+1) \cdot (2n+1)} \cdot \int_{\theta_J}^\pi P_m(\cos\theta) \cdot P_n(\cos\theta) \cdot \sin\theta \, d\theta . \tag{3.11}$$

They form the elements of the fully occupied matrices $\mathbf{Y}_<$ and $\mathbf{Y}_>$. For the sum of these matrices

$$\mathbf{E} = \mathbf{Y}_< + \mathbf{Y}_> \tag{3.12}$$

must hold with $\mathbf{E}$ representing the unit matrix. Our well-known T-matrix relation

$$\vec{c}^{\,tp} = \mathbf{T_{hs}} \cdot \vec{d}^{\,tp} , \tag{3.13}$$

follows from (3.8) after some algebra with $\mathbf{T_{hs}}$ now given by

$$\mathbf{T_{hs}} = -\left[\mathbf{Y}_< \cdot \mathbf{D_{h'}} + \mathbf{Y}_> \cdot \mathbf{D_h}\right]^{-1} \cdot \left[\mathbf{Y}_< \cdot \mathbf{D_{j'}} + \mathbf{Y}_> \cdot \mathbf{D_j}\right] \tag{3.14}$$

in matrix notation. Note that $\vec{c}^{\,tp}$ and $\vec{d}^{\,tp}$ are the column vectors with the expansion coefficients of the scattered and incident fields as components. $\mathbf{D_h}$, $\mathbf{D_j}$, $\mathbf{D_{h'}}$ und $\mathbf{D_{j'}}$ are diagonal matrices with elements

$$\mathbf{D_h} = diag[h_n^{(1)}(\beta)] \quad n = 0, \ldots, ncut \tag{3.15}$$
$$\mathbf{D_j} = diag[j_n(\beta)] \quad n = 0, \ldots, ncut \tag{3.16}$$
$$\mathbf{D_{h'}} = diag\{[h_n^{(1)}(\beta)]'\} \quad n = 0, \ldots, ncut \tag{3.17}$$
$$\mathbf{D_{j'}} = diag[j_n'(\beta)] \quad n = 0, \ldots, ncut . \tag{3.18}$$

Note that all the dashes on the Bessel and Hankel functions denote again the first derivative of these functions with respect to their arguments. From (3.10), (3.11) and from the orthonormality relation of the spherical harmonics it becomes clear that the T-matrix (3.14) of the h-s Janus sphere contains the T-matrix of the sound soft and sound hard sphere as a limiting case if $\theta_J = 0$ or $\theta_J = \pi$ is chosen.

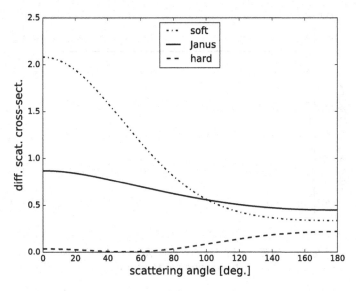

Fig. 3.2 Differential scattering cross-section of a h-s Janus sphere in axisymmetric orientation. Parameters: radius $a = 1.0$ mm, size parameter $\beta = 1.0$, splitting angle $\theta_J = 90.0°$, truncation parameter $ncut = 117$ (sound soft/hard: $ncut = 6!$). Total scattering cross-sections: $\sigma_{tot} = 7.616$ (Janus), $\sigma_{tot} = 10.626$ (sound soft), $\sigma_{tot} = 1.01$ (sound hard)

The differential and total scattering cross-sections of the h-s Janus sphere at a size parameter of $\beta = 1$ together with the corresponding results of the two limiting cases are presented in Fig. 3.2. Note that the Python program *janus_axial.py* was used for the Janus sphere. The most striking difference is the fact that the Janus sphere at such a low size parameter—although its differential scattering cross-section behaves quite boring—requires a truncation parameter of $ncut = 117$! Contrary, remember that for the two limiting cases of the sound soft and sound hard sphere at the same size parameter a truncation parameter of $ncut = 6$ was sufficient to obtain already a convergent result. We will therefore use this example to take a closer look at the convergence behavior of this type of Janus spheres and to derive an appropriate convergence strategy. It will turn out that it differs from that one mentioned at the beginning of Sect. 2.5 for the limiting cases of spheres with uniform boundary conditions. To this end, let us consider the dependence of the total scattering cross-sections on the truncation parameter $ncut$.

The total scattering cross-section as a function of $ncut$ for the limiting case of the sound soft sphere is presented in Fig. 3.3. We can observe a simple uniform behavior, and convergence is reached quickly. The same happens for the limiting case of the sound hard sphere. In contrast, the corresponding result for the Janus sphere (the computation was performed with the Python program *janus_spheres_ncut.py*) exhibits a much more complicate behavior, as it can be seen from Figs. 3.4 and 3.5. Starting from the highest peak at $ncut = 7$ we can identify three windows of conver-

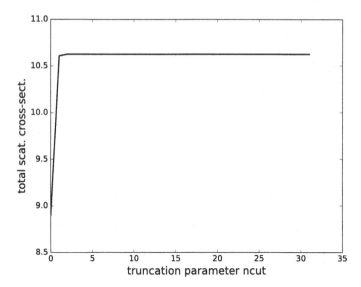

Fig. 3.3 Total scattering cross-section of the sound soft sphere of Fig. 2.1 as a function of the truncation parameter *ncut*

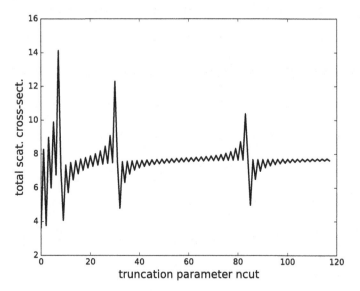

Fig. 3.4 Total scattering cross-section of the Janus sphere of Fig. 3.2 as a function of the truncation parameter *ncut*

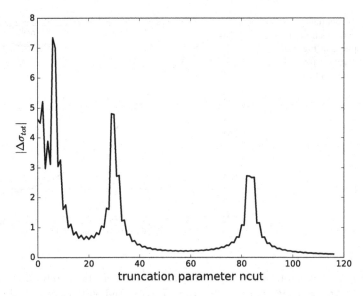

Fig. 3.5 Absolute difference of the total scattering cross-sections of the Janus sphere of Fig. 3.2 calculated at two successive truncation parameters *ncut* and *ncut* + 1

gence with strong oscillations. These oscillations decrease up to the center of each window (where the differences between two successive computations are smallest) and increase again beyond this point. Moreover, the smallest difference in the center of each window decreases from window to window while the windows themselves get broader. Regarding our example, only in the third window at *ncut* = 117 the differences between the total scattering cross-sections of two successive computations are approaching 1%. This was the threshold that was used in these computations. To determine the appropriate truncation parameter for a certain configuration of a h-s Janus sphere it is therefore recommended to proceed as follows: Start with the computation of the total scattering cross-section for an increasing number of *ncut* to identify the convergence windows. The lower of the two successive values with the smallest difference in each window should then be used as a starting point to apply the criterion for the differential scattering cross-sections of two successive computations mentioned in Sect. 2.5. The two thresholds that have to be predefined in this procedure are dependent on the application one has in mind, of course. But is it really necessary to use the truncation parameter *ncut* = 117, determined in this way, in every computation that is related to this special configuration of a Janus sphere? The answer to this question has a major impact on the numerical effort if this Janus sphere is rotated in the laboratory frame, for example. And this is exactly what we intend to discuss in the next subsection. Fortunately, this question can often be answered with "no"! In Fig. 3.6, the results for the differential scattering cross-sections of the Janus sphere of Fig. 3.2 for three different truncation parameters are depicted. The two lower truncation parameters are chosen such that they end up

Fig. 3.6 Differential
scattering cross-section of
the Janus sphere of Fig. 3.2
for three different values of
the truncation parameter

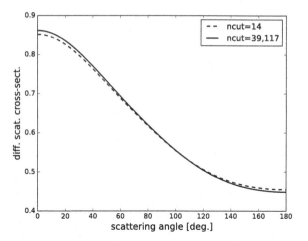

with nearly the same total scattering cross-section we have obtained for $ncut = 117$
($ncut = 14$: $\sigma_{tot} = 7.596$, $ncut = 39$: $\sigma_{tot} = 7.614$). And only if $ncut = 14$ is used
we can observe small differences in the differential scattering cross-sections. A com-
parison of the two results we obtained with $ncut = 39$ and $ncut = 117$ shows that
the numbers for the differential scattering cross-sections are identical up to two dig-
its after the decimal point. Any differences are therefore only hard to see in this
figure. But it should be clearly emphasized that this possibility does not release us
from the initial procedure to find the value for $ncut$ with a difference in the total
scattering cross-sections for two successive computations that is appropriate for a
certain application. The numerical effort can be reduced only afterwards by identi-
fying lower truncation parameters with total scattering cross-sections comparable to
that one determined in the previous step.

Beside the procedure described above it is also of some advantage to look at
the fulfillment of the boundary conditions on the two surface regions, not at least
to gain more convidence into the obtained scattering results. Since the boundary
conditions do not apply in the far-field, we have to use the series expansions (1.67)
and (1.69) for the scattered and incident field with expansion coefficients c_n of the
scattered field calculated from the T-matrix (3.14) and relation (3.13). These series
expansions must be inserted into conditions (2.1) and (2.21). The results for the
Janus sphere of Fig. 3.2 if using $ncut = 117$ are presented in Figs. 3.7 and 3.8. We
can observe the typical behavior of an oscillating approximation around zero for the
homogeneous Dirichlet and Neumann condition in the respective subregion. Such
a behavior is well-known from the least-squares approximation of a step function
by a Fourier series (see [1], for example). If we perform the same computation with
the truncation parameter of $ncut = 14$ we have to state much larger differences in
the oscillations around zero than we have found between the differential scattering
cross-sections (see Figs. 3.9 and 3.10). This confirms the known fact that it requires

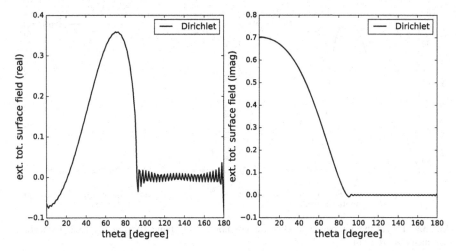

Fig. 3.7 Test of the homogeneous Dirichlet condition of the Janus sphere of Fig. 3.2. Truncation parameter: $ncut = 117$. The computation was performed with the Python program $h\_s\_axial\_surface.py$

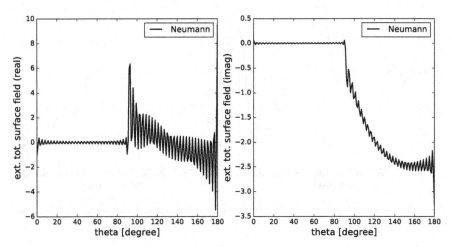

Fig. 3.8 Test of the homogeneous Neumann condition of the Janus sphere of Fig. 3.2. Truncation parameter: $ncut = 117$. The computation was performed with the Python program $h\_s\_axial\_surface.py$

a higher numerical effort to achieve a sufficiently convergent result in the near field of the scatterer than it is required in its far field (see [2], for example). The results for the total and differential scattering cross-sections and the boundary conditions of the Janus sphere of Fig. 3.2 but at a higher size parameter of $\beta = 6.0$ are depicted in Figs. 3.11, 3.12, 3.13 and 3.14.

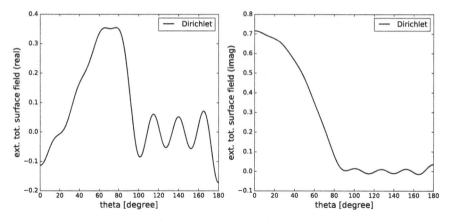

Fig. 3.9 Test of the homogeneous Dirichlet condition of the Janus sphere of Fig. 3.2. Truncation parameter: $ncut = 14$

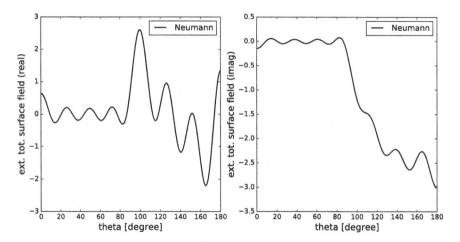

Fig. 3.10 Test of the homogeneous Neumann condition of the Janus sphere of Fig. 3.2. Truncation parameter: $ncut = 14$

It is also interesting to see what happens if we reduce one of the subregion and enlarge the other one correspondingly. In this way we are approaching the limiting case of the corresponding sphere with homogeneous boundary conditions on its whole surface. If choosing $\theta_j = 30°$, for example, we reduce the subregion that applies to the boundary condition of the sound hard sphere. That is, we should approach the limiting situation of the corresponding sound soft sphere. It is really like this, as one can see from Fig. 3.15. It can moreover be seen from Fig. 3.16 that the dependence of the total scattering cross-section on the truncation parameter $ncut$ is less dramatic than in our foregoing example. On the other hand, if choosing $\theta_j = 150°$ we do not approach the differential scattering cross-section of the sound hard sphere.

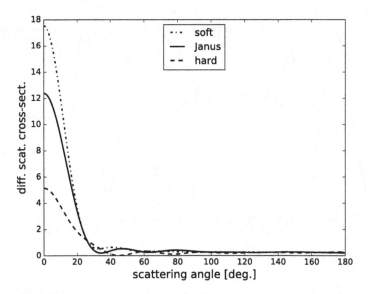

Fig. 3.11 Differential scattering cross-section of a h-s Janus sphere in axisymmetric orientation. Parameters: radius $a = 1.0$ mm, size parameter $\beta = 6.0$, splitting angle $\theta_J = 90.0°$, truncation parameter $ncut = 66$ (sound soft/hard: $ncut = 11!$). Total scattering cross-sections: $\sigma_{tot} = 7.071$ (Janus), $\sigma_{tot} = 7.983$ (sound soft), $\sigma_{tot} = 4.349$ (sound hard)

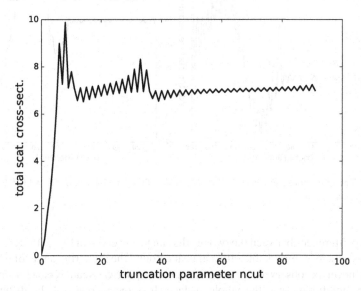

Fig. 3.12 Total scattering cross-section of the Janus sphere of Fig. 3.11 as a function of the truncation parameter $ncut$

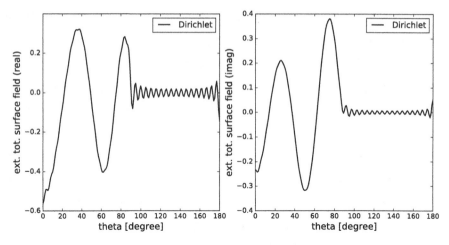

Fig. 3.13 Test of the homogeneous Dirichlet condition of the Janus sphere of Fig. 3.11. Truncation parameter: $ncut = 70$

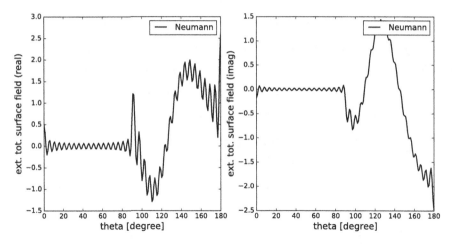

Fig. 3.14 Test of the homogeneous Neumann condition of the Janus sphere of Fig. 3.11. Truncation parameter: $ncut = 70$

Essential differences between the two results can be observed in Fig. 3.17. Regarding this Janus sphere, a remarkable effect appears around the scattering angle of $\theta = 40°$ that can never be observed for spherical or nonspherical scatterers with uniform boundary conditions over the whole surface. It is the increase of the differential scattering cross-section around this angle that goes beyond the value of the forward direction! Since this appears as a ring of a stronger intensity with an opening angle of 40° if looking from above we will call this effect the "halo effect" of Janus spheres. However, it differs from the well-known halo effect of hexagonal ice crystals in that the intensity of the Janus halo is stronger than the intensity in forward direction.

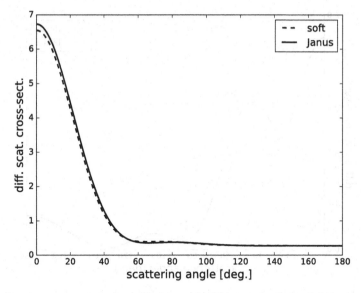

Fig. 3.15 Differential scattering cross-section of a h-s Janus sphere in axisymmetric orientation. Parameters: radius $a = 1.0$ mm, size parameter $\beta = 3.0$, splitting angle $\theta_J = 30.0°$, truncation parameter $ncut = 31$ (sound soft: $ncut = 8!$). Total scattering cross-sections: $\sigma_{tot} = 9.061$ (Janus), $\sigma_{tot} = 8.813$ (sound soft)

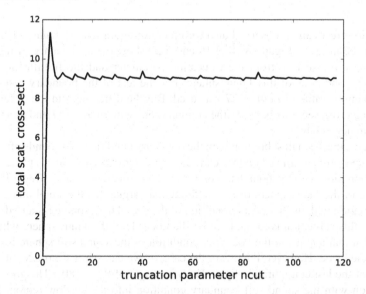

Fig. 3.16 Total scattering cross-section of the Janus sphere of Fig. 3.15 as a function of the truncation parameter $ncut$

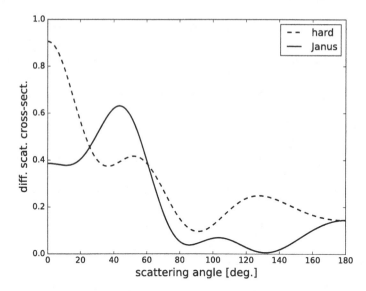

Fig. 3.17 Differential scattering cross-section of a h-s Janus sphere in axisymmetric orientation. Parameters: radius $a = 1.0$ mm, size parameter $\beta = 3.0$, splitting angle $\theta_J = 150.0°$, truncation parameter $ncut = 57$ (sound hard: $ncut = 8$!). Total scattering cross-sections: $\sigma_{tot} = 2.385$ (Janus), $\sigma_{tot} = 3.247$ (sound hard)

This halo effect can be observed also for other configurations of Janus spheres, as we will see later on. Figures 3.18, 3.19 and 3.20 show the dependence of the total scattering cross-section from the truncation parameter and the fulfillment of the boundary conditions of this Janus sphere. For the test of the boundary conditions the maximum value of $ncut = 77$ was used. But the differences in the differential scattering cross-sections between the computations with $ncut = 57$ and $ncut = 77$ are again neglectable.

It seems that a small subregion with the boundary condition of a sound soft sphere has a much stronger impact on the scattering behavior of the h-s Janus sphere than a small subregion with the boundary condition of a sound hard sphere. Regarding the behavior of the Janus sphere of Fig. 3.15 one may argue that the small influence of the subregion with the boundary condition of the sound hard sphere is caused by the fact that this subregion is located on the shadow side of the Janus sphere while the subregion that applies to the boundary condition of the sound soft sphere is on the illuminated side. However, this is not the case, as we can see if we rotate the Janus sphere of the last configuration by an Eulerian angle of $\theta_p = 180°$. This moves the subregion with the sound soft boundary condition into the shadow region. In the following subsection we will therefore derive the T-matrix of the Janus sphere of the h-s type if arbitrarily rotated in the laboratory frame. This will allow us moreover to discuss the reciprocity condition and its usage as a further criterion for the physical meaningfulness and accuracy of the obtained scattering results.

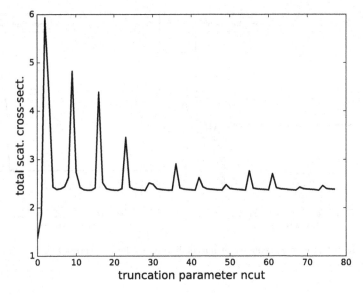

Fig. 3.18 Total scattering cross-section of the Janus sphere of Fig. 3.17 as a function of the truncation parameter *ncut*

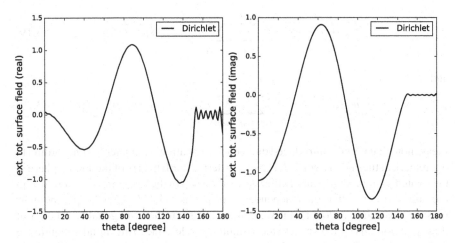

Fig. 3.19 Test of the homogeneous Dirichlet condition of the Janus sphere of Fig. 3.17. Truncation parameter: $ncut = 77$

3.1.2 Arbitrary Rotation

The derivation of the T-matrix of the rotated Janus sphere requires only little additional effort since we can start with the same procedure used in Sect. 2.3.1. We transform the incident plane wave according to (1.71)/(1.72) and use expansion (2.44)

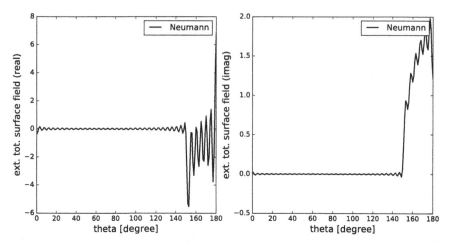

Fig. 3.20 Test of the homogeneous Neumann condition of the Janus sphere of Fig. 3.17. Truncation parameter: $ncut = 77$

for the scattered field in the rotated system. That is, we have

$$u_{\text{inc}}(k_0\bar{r}, \bar{\theta}, \bar{\phi}) = \sum_{\bar{l}=-lcut}^{lcut} \sum_{n=|\bar{l}|}^{ncut} \widehat{d}_{\bar{l}n} \cdot \psi_{\bar{l}n}(k_0\bar{r}, \bar{\theta}, \bar{\phi}) \qquad (3.19)$$

and

$$u_s(k_0\bar{r}, \bar{\theta}, \bar{\phi}) = \sum_{\bar{l}=-lcut}^{lcut} \sum_{n=|\bar{l}|}^{ncut} \widehat{c}_{\bar{l}n} \cdot \varphi_{\bar{l}n}(k_0\bar{r}, \bar{\theta}, \bar{\phi}) \ . \qquad (3.20)$$

Please note that we now introduced the second truncation parameter $lcut$ according to our remark at the end of Sect. 2.3.1! Now, to derive the T-matrix of the Janus sphere in the rotated system, we can apply the same steps used in the foregoing subsection with the only difference that the l-dependence of the associated Legendre polynomials must be taken into account. But we can benefit from the orthogonality relation of these polynomials with respect to the azimuthal modes, due to the initially mentioned restriction to rotational symmetric subregions. Looking at (3.14) we eventually obtain the T-matrix

$$\mathbf{T}_{\mathbf{hs}}^{(\bar{l})} = - \left[\mathbf{Y}_<^{(\bar{l})} \cdot \mathbf{D}_{h'} + \mathbf{Y}_>^{(\bar{l})} \cdot \mathbf{D}_h \right]^{-1} \cdot \left[\mathbf{Y}_<^{(\bar{l})} \cdot \mathbf{D}_{j'} + \mathbf{Y}_>^{(\bar{l})} \cdot \mathbf{D}_j \right] \qquad (3.21)$$

of the h-s Janus sphere in the rotated system with the two matrices $\mathbf{Y}_<^{(\bar{i})}$ and $\mathbf{Y}_>^{(\bar{i})}$ given by

$$
\left[Y_<^{(\bar{i})}\right]_{mn} = \frac{1}{2} \cdot \sqrt{\frac{(2m+1) \cdot (2n+1) \cdot (m-\bar{i})! \cdot (n-\bar{i})!}{(m+\bar{i})! \cdot (n+\bar{i})!}} \cdot
$$

$$
\int_0^{\theta_J} P_m^{\bar{i}}(\cos\theta) \cdot P_n^{\bar{i}}(\cos\theta) \cdot \sin\theta \, d\theta \; ; \quad \bar{i} = -lcut, \ldots, lcut
$$

$$
m, n = |\bar{i}|, \ldots, ncut \qquad (3.22)
$$

and

$$
\left[Y_>^{(\bar{i})}\right]_{mn} = \frac{1}{2} \cdot \sqrt{\frac{(2m+1) \cdot (2n+1) \cdot (m-\bar{i})! \cdot (n-\bar{i})!}{(m+\bar{i})! \cdot (n+\bar{i})!}} \cdot
$$

$$
\int_{\theta_J}^{\pi} P_m^{\bar{i}}(\cos\theta) \cdot P_n^{\bar{i}}(\cos\theta) \cdot \sin\theta \, d\theta \; ; \quad \bar{i} = -lcut, \ldots, lcut
$$

$$
m, n = |\bar{i}|, \ldots, ncut \; . \qquad (3.23)
$$

$$
\mathbf{E} = \mathbf{Y}_<^{(\bar{i})} + \mathbf{Y}_>^{(\bar{i})} \qquad (3.24)
$$

holds again for the sum of these matrices. The coefficients $\widehat{c}_{\bar{i}n}$ in expansion (3.20) are then given by

$$
\widehat{c}_{\bar{i}}^{\,tp} = \mathbf{T}_{hs}^{(\bar{i})} \cdot \vec{d}_{\bar{i}}^{\,tp} \; ; \quad \bar{i} = -lcut, \ldots, lcut \qquad (3.25)
$$

in matrix notation. Note that $\mathbf{T}^{(\bar{i})}{}_{hs}$ is again a full but now $\bar{i}$-dependent matrix. The transformation of the scattered field (3.20) back into the laboratory frame is finally accomplished by using (1.38).

These equations are implemented in the Python program *janus_rotated.py*. To determine the two truncation parameters *ncut* and *lcut* for a certain orientation (α, θ_p) of the Janus sphere the following way has proved favorable: First we choose $\alpha = \theta_p = 0°$ (this is the axisymmetric orientation) and determine *ncut* as discussed in the foregoing subsection. Program *janus_axial.py* can be used for this purpose. This value of *ncut* is used afterwards to determine *lcut* for the given values (α, θ_p) by comparing two successive computations for *lcut* and *lcut* + 1. The computation is stopped if the differences in the differential scattering cross-sections are again below a certain threshold in 80% of the scattering angles. We have already successfully applied such a procedure in the light scattering analysis of nonspherical but rotational symmetric scatterers (see [3], Chap. 9 therein).

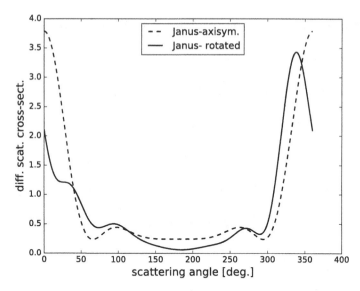

Fig. 3.21 Differential scattering cross-sections of an axisymmetrically oriented and rotated h-s Janus sphere. Parameters: radius $a = 1.0$ mm, size parameter $\beta = 3.0$, splitting angle $\theta_J = 90.0°$, Eulerian angles of rotation $(\alpha = 0°, \theta_p = 90°)$, truncation parameters $ncut = 38, lcut = 5$. Total scattering cross-sections: $\sigma_{\text{tot}} = 5.398$ (rotated), $\sigma_{\text{tot}} = 7.370$ (axisymmetric)

The differential and total scattering cross-sections of a h-s Janus sphere in axisymmetric orientation and if rotated by an angle of $\theta_p = 90°$ in the scattering plane is represented in Fig. 3.21. This example was chosen since we can observe an intensity of the rotated configuration at a scattering angle of $\theta = 338°$ that goes again beyond the forward direction. But this effect cannot be observed if looking at the axisymmetric orientation of this Janus sphere. It disappears also if the orientation $(\alpha = 90°, \theta_p = 90°)$ is chosen, as demonstrated with Fig. 3.22. That is, if we consider an ensemble of such Janus spheres that is assumed to scatter independently, and that is randomly oriented with respect to the Eulerian angle α but at fixed $\theta_p = 90°$, then we would see again the halo ring if looking from above on this ensemble! But we can observe another difference in the differential scattering cross-sections of this Janus sphere. Regarding the rotated Janus sphere of Fig. 3.21, we get an asymmetric scattering behavior with respect to the z-axis of the scattering plane. On the other hand, the axisymmetric orientation and the orientation used in Fig. 3.22 result in a symmetric behavior, as one can expect. This example emphasizes also the importance of the additionally introduced truncation parameter $lcut$. We can obviously choose $lcut$ much smaller than $ncut$! This results in a drastic reduction of the numerical effort although it goes still well beyond the numerical effort that is required for the corresponding sphere with uniform boundary conditions over its whole surface. The total scattering cross-sections of the rotated Janus spheres of Figs. 3.21 and 3.22

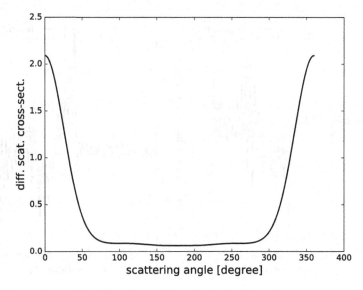

Fig. 3.22 Differential scattering cross-sections of the h-s Janus sphere of Fig. 3.21 but if rotated by the Eulerian angles ($\alpha = 90°$, $\theta_p = 90°$). Other parameters as in Fig. 3.21. Total scattering cross-section: $\sigma_{tot} = 5.398$

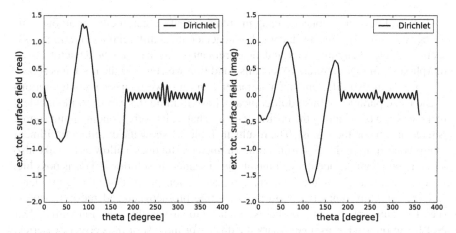

Fig. 3.23 Test of the Dirichlet condition of the h-s Janus sphere of Fig. 3.21. Truncation parameter as in Fig. 3.21

show moreover that this scattering quantity is independent of the Eulerian angle α! The approximation of the Dirichlet and Neumann boundary conditions of the rotated Janus sphere of Fig. 3.21 is depicted in Figs. 3.23 and 3.24.

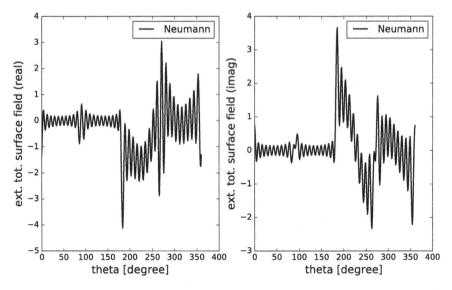

Fig. 3.24 Test of the Neumann condition of the h-s Janus sphere of Fig. 3.21. Truncation parameter as in Fig. 3.21

In Sect. 1.4 we mentioned already the requirement of reciprocity for all our scattering problems. This means that we should get identical differential scattering cross-sections if interchanging source and measurement points in the far-field but for a fixed morphology of the scatterer. To test this condition we consider the h-s Janus sphere of Fig. 3.21 at two different size parameters, and for two reciprocal configurations at each size parameter. In so doing, please, have in mind that we fixed the incident plane wave to travel along the positive z-axis. That is, its corresponding point source is fixed at $-\infty$ on the z-axis!. The results in Table 3.1 show the excellent fulfillment of this condition for the used truncation parameters. But does it allow us to make also a statement about the accuracy of the obtained scattering solutions? This is possible, for example, if applying the T-matrix approach to nonspherical scattering problems (see [3, 4]). Unfortunately, for all the scattering problems considered in this book (i.e., for Janus spheres and bispheres) we have to answer this question with a clear **No!** It can only serve as a necessary condition for the test of the correct numerical implementation of the T-matrix approach, although the importance of this possibility should not be underestimated! This can be seen if we reduce the truncation parameters used in Table 3.1 in such a drastic way that the T-matrix approach produces obviously nonsens. But if the same truncation parameters are used for the reciprocal configurations we still get an excellent agreement between the respective differential scattering cross-sections. Regarding the two reciprocal configurations with ($\alpha = 0°, \theta_p = 0°$) and ($\alpha = 0°, \theta_p = 90°$) (this corresponds to the first two columns in Table 3.1) it moreover turns out that the differential scattering cross-sections of the configuration ($\alpha = 0°, \theta_p = 90°$) are independent of the number of $lcut$. But it is interesting that this holds only for the reciprocal scattering angle θ! The differential scattering

Table 3.1 Test of the reciprocity condition. The numbers represent the differential scattering cross-sections at the reciprocal scattering angles θ

Config.	$\alpha = 0°$ $\theta_p = 0°$ $\theta = 90°$	$\alpha = 0°$ $\theta_p = 90°$ $\theta = 270°$	$\alpha = 0°$ $\theta_p = 45°$ $\theta = 60°$	$\alpha = 0°$ $\theta_p = 165°$ $\theta = 300°$
$\beta = 1.0$ ($ncut = 39, lcut = 3$)	0.586	0.586	0.557	0.557
$\beta = 3.0$ ($ncut = 38, lcut = 5$)	0.426	0.426	0.217	0.217

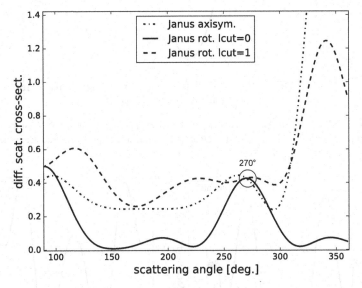

Fig. 3.25 Differential scattering cross-sections of the h-s Janus sphere of Fig. 3.21 in axisymmetric orientation and if rotated by the Eulerian angles ($\alpha = 0°$, $\theta_p = 90°$) but for the two lower values of $lcut$

cross-sections at the other scattering angles differ for different values of $lcut$. This can be seen from Fig. 3.25 where we have performed the same computations as in Fig. 3.21 but for two lower values of $lcut$. All three curves have a common point at the scattering angle of $\theta = 270°$ although the two curves with $lcut = 0, 1$ do not agree with the final result of the rotated Janus sphere we obtained with $lcut = 5$.

In the last example of this section we will answer the question if the influence of a small sound soft subregion of a h-s Janus sphere has a minor impact on the the differential scattering cross-section if rotated into the shadow region of this Janus sphere (see the remark at the end of Sect. 3.3.1, and Fig. 3.17). But the influence, however, is still significant, as it can be seen from Fig. 3.26. The corresponding approximations of the boundary conditions are depicted in Figs. 3.27 and 3.28.

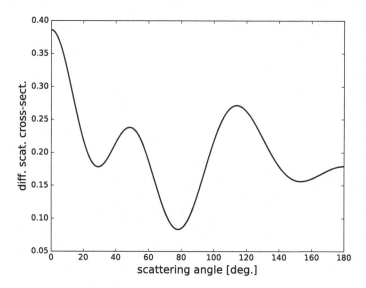

Fig. 3.26 Differential scattering cross-sections of the h-s Janus sphere of Fig. 3.17 but if rotated by the Eulerian angles ($\alpha = 0°$, $\theta_p = 180°$). Other parameters as in Fig. 3.17. Total scattering cross-section: $\sigma_{\text{tot}} = 2.385$

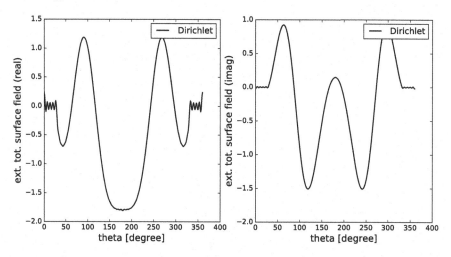

Fig. 3.27 Test of the Dirichlet condition of the h-s Janus sphere of Fig. 3.26

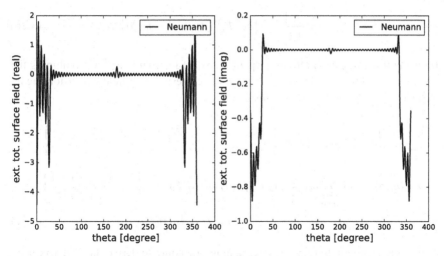

Fig. 3.28 Test of the Neumann condition of the h-s Janus sphere of Fig. 3.26

3.2 Janus Sphere of s-p Type

3.2.1 Axisymmetric Orientation

We start again with the simpler situation of the axisymmetric orientation of this type
of Janus sphere. The dashed subregion in Fig. 3.1 is now assigned to the boundary
condition (2.1) of the sound soft sphere, while the other subregion applies to the
transmission conditions (2.25) and (2.26) of the sound penetrable sphere. This combination corresponds to the combination of the boundary conditions of a dielectric
and ideal metallic sphere in the electromagnetic case. The derivation of the T-matrix
requires a slight modification of the approach we described in Sect. 3.1.1. The following procedure is similar to the procedure known from the analysis of planar antennas
and microwave structures, as described in [6, 7], for example. It is even the more
remarkable that it has not been applied yet to scattering problems on Janus spheres.

To be able to apply the boundary condition (3.1) we first have to use condition
(2.25) to express the unknown coefficients g_n of expansion (1.66) of the internal
field in terms of the unknown expansion coefficients c_n of the scattered field and the
known expansion coefficients d_n of the incident plane wave. This intermediate step
follows from the fact that the Dirichlet condition must hold on the whole surface of
this Janus sphere. I.e., it must also hold if we are approaching the dashed subregion
in Fig. 3.1 from inside! This is enforced (on the level of an approximation in terms
of the employed series expansions) by using (2.25) to eliminate g_n. Because of the
axisymmetric orientation only the azimuthal mode $l = 0$ has to be considered. We
thus get

$$g_n = \frac{1}{j_n(\beta_p)} \cdot \left[c_n \cdot h_n^{(1)}(\beta) + d_n \cdot j_n(\beta) \right] . \tag{3.26}$$

Inserting this expression into the expansion of the internal field, and combining the boundary conditions (2.1) and (2.26) and the corresponding field expansions again in boundary condition (3.1) now provides

$$\sum_{n=0}^{ncut} c_n \cdot \left\{ h_n^{(1)}(\beta) \cdot f_<(\theta) + \left\{ \left[h_n^{(1)}(\beta) \right]' - \frac{1}{\kappa} \cdot \frac{h_n^{(1)}(\beta) \cdot j_n'(\beta_p)}{j_n(\beta_p)} \right\} \cdot f_>(\theta) \right\} \cdot$$

$$Y_{0n}(\theta, \phi) = -\sum_{n=0}^{ncut} d_n \cdot \left\{ j_n(\beta) \cdot f_<(\theta) + \left[j_n'(\beta) - \frac{1}{\kappa} \cdot \frac{j_n(\beta) \cdot j_n'(\beta_p)}{j_n(\beta_p)} \right] \cdot \right.$$

$$\left. f_>(\theta) \right\} \cdot Y_{0n}(\theta, \phi) . \tag{3.27}$$

This is the analogon to (3.8). The remaining procedure to derive the T-matrix from this equation is identical with the procedure that was applied subsequent to (3.8). In so doing, we end up with the T-matrix

$$\mathbf{T_{sp}} = - \left[\mathbf{Y}_< \cdot \mathbf{D_h} + \mathbf{Y}_> \cdot \mathbf{D_c'} \right]^{-1} \cdot \left[\mathbf{Y}_< \cdot \mathbf{D_j} + \mathbf{Y}_> \cdot \mathbf{D_d'} \right] \tag{3.28}$$

of the Janus sphere of s-p type. $\mathbf{D_h}$ and $\mathbf{D_j}$ are identical with (3.15) and (3.16), while the diagonal matrices $\mathbf{D_c'}$ and $\mathbf{D_d'}$ are given by

$$\mathbf{D_c'} : \left[h_n^{(1)}(\beta) \right]' - \frac{1}{\kappa} \cdot \frac{h_n^{(1)}(\beta) \cdot j_n'(\beta_p)}{j_n(\beta_p)} \quad n = 0, \dots, ncut \tag{3.29}$$

$$\mathbf{D_d'} : j_n'(\beta) - \frac{1}{\kappa} \cdot \frac{j_n(\beta) \cdot j_n'(\beta_p)}{j_n(\beta_p)} \quad n = 0, \dots, ncut . \tag{3.30}$$

The T-matrices of the sound soft and sound penetrable sphere are again limiting cases of (3.1) if the splitting angle $\theta_J = 180°$ or $\theta_J = 0°$ is used. The expansion coefficients of the scattered field are computed from

$$\vec{c}^{tp} = \mathbf{T_{sp}} \cdot \vec{d}^{tp} . \tag{3.31}$$

We will now apply this T-matrix to solve the scattering problem on s-p Janus spheres for the same configurations (i.e., the configuration with respect to the size parameter and orientation) used in the previous section. Beside providing results for intercomparison purposes, with the following examples we intend also to demonstrate the differences in the behavior of the different types of Janus spheres. The differential and total scattering cross-sections of the configuration used in Fig. 3.2 but if applied to the s-p Janus sphere are presented in Fig. 3.29. The truncation param-

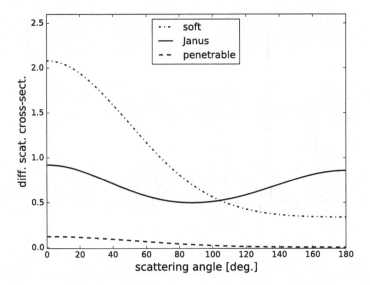

Fig. 3.29 Differential scattering cross-section of a s-p Janus sphere in axisymmetric orientation. Parameters: radius $a = 1.0$ mm, size parameter $\beta = 1.0$, $\kappa_k = \rho_p = 1.5$, splitting angle $\theta_J = 90.0°$, truncation parameter $ncut = 33$. Total scattering cross-sections: $\sigma_{tot} = 7.764$ (Janus), $\sigma_{tot} = 10.626$ (sound soft), $\sigma_{tot} = 0.5276$ (sound penetrable)

eter $ncut$ is determined in the same way that was already used for the h-s Janus sphere in axisymmetric orientation. Note that the Python program *janus_axial.py* was again used for this type of Janus spheres. Comparing this result with the differential scattering cross-section of the h-s Janus sphere of Fig. 3.2 we can observe qualitative differences in the side- and backscattering region. There is now a minimum at a scattering angle of $\theta = 90°$ and an increased backscattering intensity that goes well beyond the intensity of the sound soft sphere. However, the differences in the total scattering cross-sections are less pronounced. The dependence of the total scattering cross-section on the truncation parameter $ncut$ is shown in Figs. 3.30 and 3.31. It shows again the characteristic behavior we obtained already for the h-s Janus sphere. The results of the test of the Dirichlet condition on the surface of the upper subregion for two different truncation parameters $ncut$ are depicted in Figs. 3.32 and 3.33. Although the result obtained with $ncut = 101$ shows a much better approximation of this condition if compared to $ncut = 33$, no differences between the differential scattering cross-sections for these two values can be observed. The test of the two parts of the penetrable boundary conditions is shown in Figs. 3.34 and 3.35. Note that the Dirichlet condition is now fulfilled within double precision accuracy on the whole surface since forced by relation (3.26)!

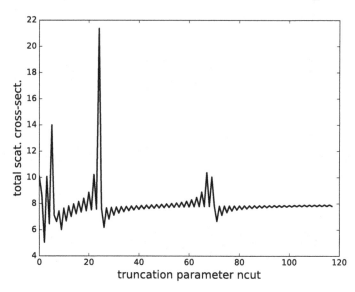

Fig. 3.30 Total scattering cross-section of the Janus sphere of Fig. 3.29 as a function of the truncation parameter *ncut*

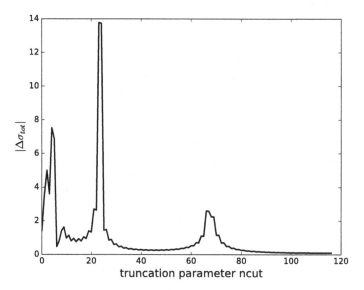

Fig. 3.31 Absolute difference of the total scattering cross-sections of the Janus sphere of Fig. 3.29 calculated at two successive truncation parameters *ncut* and *ncut* + 1

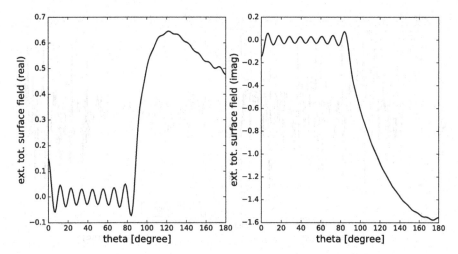

Fig. 3.32 Test of the homogeneous Dirichlet condition of the Janus sphere of Fig. 3.29. Truncation parameter: $ncut = 33$

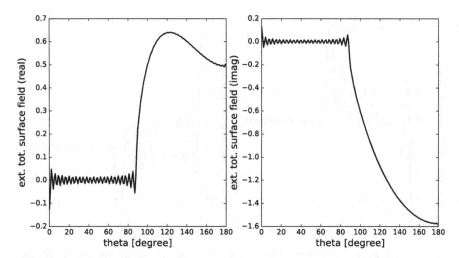

Fig. 3.33 Test of the homogeneous Dirichlet condition of the Janus sphere of Fig. 3.29. Truncation parameter: $ncut = 101$

What happens if we increase the size parameter of this Janus sphere to $\beta = 6$? This can be seen from Fig. 3.36. The differences in the backscattering region become even more pronounced. And, in contrast to the situation we found in Figs. 3.11 and 3.29, there is now an intensity of the s-p Janus sphere in forward direction that is below the corresponding intensities of the limiting spheres with uniform boundary conditions. It seems moreover that the convergence behavior of this combination of boundary conditions is less dramatic than it was found for the h-s Janus spheres.

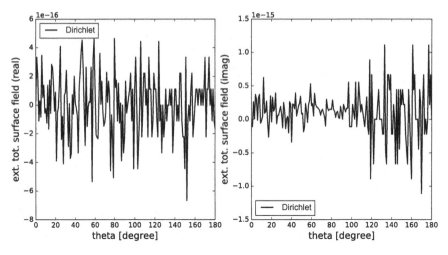

Fig. 3.34 Test of the Dirichlet condition as part of the penetrable boundary condition of the Janus sphere of Fig. 3.29. Truncation parameter: $ncut = 33$

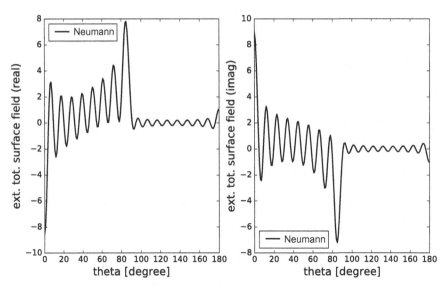

Fig. 3.35 Test of the Neumann condition as part of the penetrable boundary condition of the Janus sphere of Fig. 3.29. Truncation parameter: $ncut = 33$

While in Fig. 3.11 a truncation parameter of $ncut = 70$ was used to achieve a certain accuracy, $ncut = 36$ is needed for the s-p Janus sphere at the same size parameter, and for the same accuracy criterion. It is therefore of some worth to consider again the dependence of the total scattering cross-section on the truncation parameter $ncut$, and to compare the results obtained with the highest value of ncut with the lowest difference of two successive computations with the much lower truncation parameter

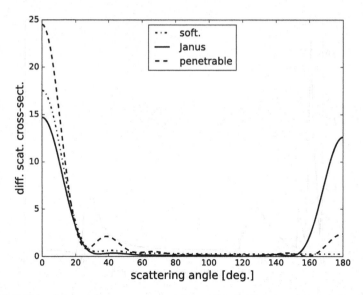

Fig. 3.36 Differential scattering cross-section of a s-p Janus sphere in axisymmetric orientation. Parameters: radius $a = 1.0$ mm, size parameter $\beta = 6.0$, $\kappa_k = \rho_p = 1.5$, splitting angle $\theta_J = 90.0°$, truncation parameter $ncut = 36$. Total scattering cross-sections: $\sigma_{tot} = 7.549$ (Janus), $\sigma_{tot} = 7.983$ (soun soft), $\sigma_{tot} = 9.771$ (sound penetrable)

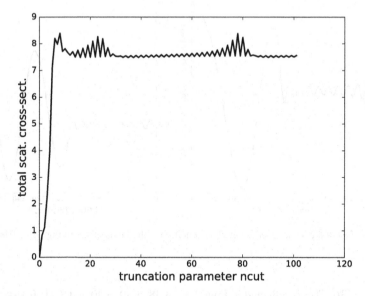

Fig. 3.37 Total scattering cross-section of the Janus sphere of Fig. 3.36 as a function of the truncation parameter $ncut$

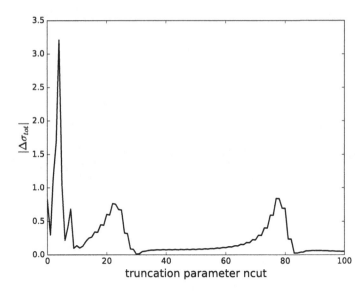

Fig. 3.38 Absolute difference of the total scattering cross-sections of the Janus sphere of Fig. 3.36 calculated at two successive truncation parameters $ncut$ and $ncut + 1$

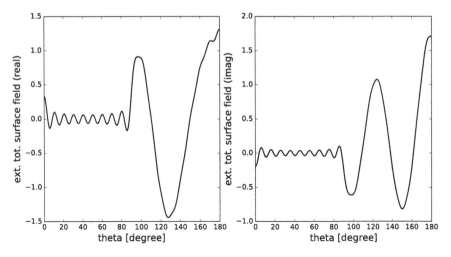

Fig. 3.39 Test of the homogeneous Dirichlet condition of the Janus sphere of Fig. 3.36. Truncation parameter: $ncut = 36$

of $ncut = 36$. This is reflected in Figs. 3.37, 3.38, 3.39, 3.40 and 3.41. It also shows up here again that the differential scattering cross-sections are nearly identical (except the small deviation at the scattering angle of $\theta = 180°$, as can be seen from Fig. 3.41) although the homogeneous Dirichlet condition on the surface of the upper subregion is much more accurate if $ncut = 101$ is used. This agrees with the behavior of all the

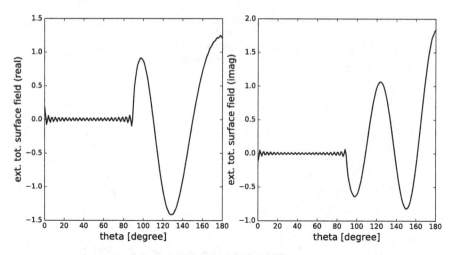

Fig. 3.40 Test of the homogeneous Dirichlet condition of the Janus sphere of Fig. 3.36. Truncation parameter: *ncut* = 101

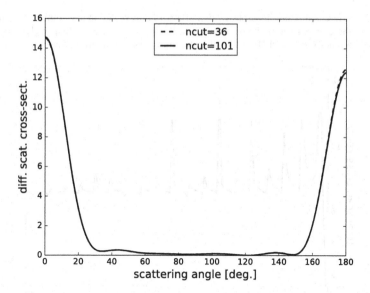

Fig. 3.41 Differential scattering cross-sections of the Janus sphere of Fig. 3.36 for two different values of the truncation parameter

previous examples and must be taken into account if one is interested in the analysis of multiple scattering processes on Janus spheres that are close to each other.

Next we will again reduce one subregion and enlarge the other one correspondingly. Choosing $\theta_J = 30°$ reduces the subregion that applies to the boundary condition of the sound soft sphere and enlarges the subregion that belongs to the sound penetrable sphere. Although the small sound soft subregion is located in the shadow

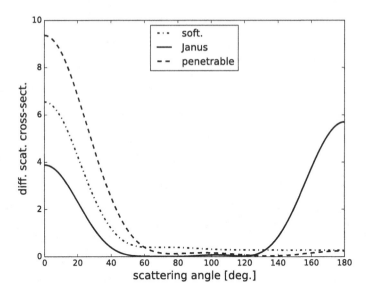

Fig. 3.42 Differential scattering cross-section of a s-p Janus sphere in axisymmetric orientation. Parameters: radius $a = 1.0\,\mathrm{mm}$, size parameter $\beta = 3.0$, $\kappa_k = \rho_p = 1.5$, splitting angle $\theta_J = 30.0°$, truncation parameter $ncut = 121(38)$. Total scattering cross-section: $\sigma_{\mathrm{tot}} = 5.695$

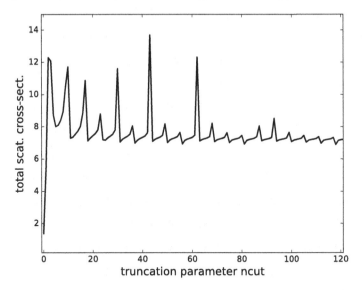

Fig. 3.43 Total scattering cross-section of the Janus sphere of Fig. 3.42 as a function of the truncation parameter $ncut$

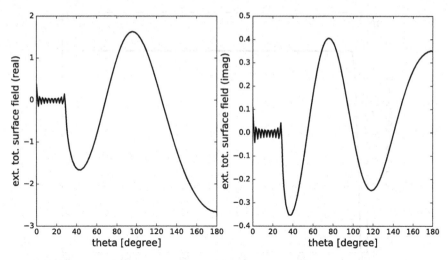

Fig. 3.44 Test of the homogeneous Dirichlet condition of the Janus sphere of Fig. 3.42. Truncation parameter: $ncut = 121$

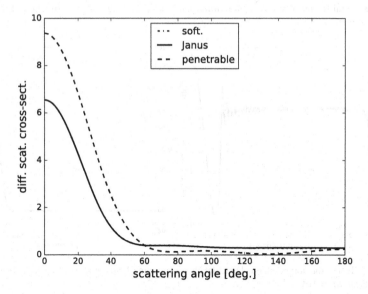

Fig. 3.45 Differential scattering cross-section of a s-p Janus sphere in axisymmetric orientation. Parameters: radius $a = 1.0\,\mathrm{mm}$, size parameter $\beta = 3.0$, $\kappa_k = \rho_p = 1.5$, splitting angle $\theta_J = 150.0°$, truncation parameter $ncut = 101(24)$. Total scattering cross-section: $\sigma_{\mathrm{tot}} = 8.821$

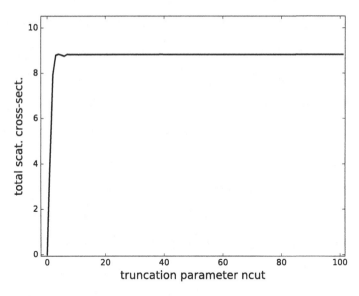

Fig. 3.46 Total scattering cross-section of the Janus sphere of Fig. 3.45 as a function of the truncation parameter *ncut*

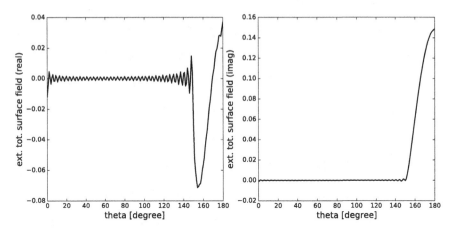

Fig. 3.47 Test of the homogeneous Dirichlet condition of the Janus sphere of Fig. 3.45. Truncation parameter: $ncut = 101$

area of the in large measure sound penetrable sphere it still has a major impact on its scattering behavior. The differential scattering cross-section is far away from the two limiting cases, as one can see from Fig. 3.42. The dependence of the total scattering cross-section on the truncation parameter is presented in Fig. 3.43 and shows strong oscillations. But using the maximum value of $ncut = 121$ or $ncut = 38$ provides no differences in both the differential and total scattering cross-sections. Figure 3.44 shows that the boundary condition of the sound soft sphere is well approached on the respective part of the surface. A total different behavior can be observed if choosing $\theta_J = 150°$. Now we have a small subregion that belongs to the sound penetrable sphere, and that is located on the illuminated side of the in large measure sound soft sphere. The results of this configuration are presented in Figs. 3.45, 3.46 and 3.47. No differences in the differential scattering cross-sections of the homogeneous sound soft sphere and the Janus sphere can be observed. And the convergence behavior is similar to that one of the homogeneous sound soft sphere so that a truncation parameter of $ncut = 24$ is already sufficient to produce a quite accurate result. This emphasizes again that the boundary condition of the sound soft sphere has a major impact on the scattering behavior of the two types of Janus spheres considered so far, even if it occupies only a small part of the boundary surface.

Acoustic or electromagnetic scattering on a flat circular disc is a scattering problem in physics that is often solved approximately with Babinet's principle and the Kirchhoff approximation. The Janus sphere of s-p type considered in this section allows us to study the scattering behavior of a bended circular disc—a sound soft "contact lense", so to speak—in a rigorous way. This is simply achieved by choosing the material parameters $\kappa_k = \rho_p = 1.0$ of the free space for the sound penetrable part of the Janus sphere. The bending and the diameter of the disc can be controlled by the radius a of the Janus sphere and the splitting angle θ_J. Increasing a allows us in principle to approach the flat circular disc. This is an alternative approach to that one published in [8]. The flat circular disc was considered in this paper as the limiting case of an oblate spheroid with an aspect ratio that tends to zero. The conventional T-matrix approach was used to solve this scattering problem at very small size parameters. The results we obtained with the T-matrix derived above for a "contact lense" are shown in Figs. 3.48, 3.49 and 3.50. Figure 3.49 is nothing but the square root of the differential scattering cross-section of Fig. 3.48 in the interval $\theta \in [0, 90°]$. It was plotted here since it can directly be compared with the results of the flat circular disc at a size parameter of 7.1 published in [9] (Fig. 5 therein). And, finally, Fig. 3.50 shows again the approximation of the homogeneous Dirichlet condition in the sound soft subregion. At the end of Sect. 2.4 we outlined a way to generalize theT-matrix approach to become applicable to nonspherical but rotational symmetric objects. This generalization—in combination with the method presented in this chapter—can also be applied to rotational symmetric objects of Janus types. An oblate spheroid of s-p type, for example, would then be another possibility to approach the scattering behavior of a flat circular disc. Would not that be a nice semester job? This requires only slight modifications of the corresponding programs provided with this book.

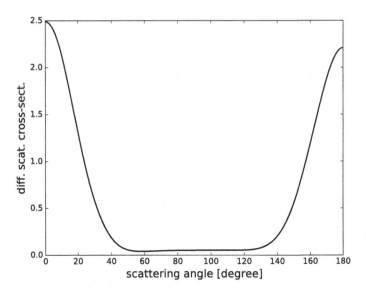

Fig. 3.48 Differential scattering cross-section of a s-p Janus sphere in axisymmetric orientation. Parameters: radius $a = 1.0$ mm, size parameter $\beta = 6.0$, $\kappa_k = \rho_p = 1.0$, splitting angle $\theta_J = 45.0°$, truncation parameter $ncut = 140\,(66)$. Total scattering cross-sections: $\sigma_{\text{tot}} = 3.155\,(3.153)$. The values in the parentheses are the alternative truncation parameter and the corresponding total scattering cross-section. Both values of $ncut$ provide nearly identical differential scattering cross-sections

3.2.2 Arbitrary Rotation

To derive the T-matrix of an arbitrarily rotated Janus sphere of s-p type is straightforward and combines the methods described in Sect. 3.1.2 and in the previous subsection. That is, expressing the expansion coefficients of the internal field in terms of the expansion coefficients of the scattered and incident field by application of the boundary condition (2.25), and the determination of the unknown expansion coefficients of the scattered field by application of the two remaining boundary conditions (2.1) and (2.26) must now be accomplished in the rotated system after the transformation of the field expansions. Looking at (3.28) the T-matrix of the s-p Janus sphere is thus given by

$$\mathbf{T}_{\mathbf{sp}}^{(\mathbf{l})} = -\left[\mathbf{Y}_{<}^{(\mathbf{l})} \cdot \mathbf{D_h} + \mathbf{Y}_{>}^{(\mathbf{l})} \cdot \mathbf{D'_c}\right]^{-1} \cdot \left[\mathbf{Y}_{<}^{(\mathbf{l})} \cdot \mathbf{D_j} + \mathbf{Y}_{>}^{(\mathbf{l})} \cdot \mathbf{D'_d}\right] \qquad (3.32)$$

in the rotated system. The elements of the two matrices $\mathbf{Y}_{<}^{(\mathbf{l})}$ and $\mathbf{Y}_{>}^{(\mathbf{l})}$ are identical with (3.22) and (3.23). Instead of (3.25) we now have

$$\vec{\tilde{c}}_{\bar{l}}^{\,tp} = \mathbf{T}_{\mathbf{sp}}^{(\bar{\mathbf{l}})} \cdot \vec{\tilde{d}}_{\bar{l}}^{\,tp} \; ; \quad \bar{l} = -lcut, \ldots, lcut \qquad (3.33)$$

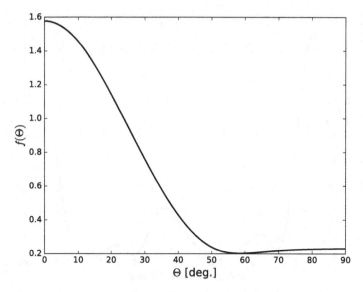

Fig. 3.49 The modulus of the scattering amplitude function for $\theta \in [0, 90°]$

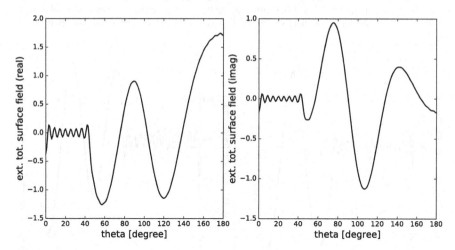

Fig. 3.50 Test of the homogeneous Dirichlet condition of the Janus sphere of Fig. 3.48. Truncation parameter: $ncut = 66$

for the expansion coefficients of the scattered field in the rotated system. And, finally, the transformation of the scattered field (3.20) back into the laboratory frame must again be accomplished by using (1.38).

Let us apply this T-matrix to solve the scattering problem that was already considered in Fig. 3.21, but now for the s-p Janus sphere. Python program *janus_rotated.py* is used for this purpose. The cross-sections and the approximation of the homogeneous Dirichlet condition are shown in Figs. 3.51 and 3.52. We can observe an

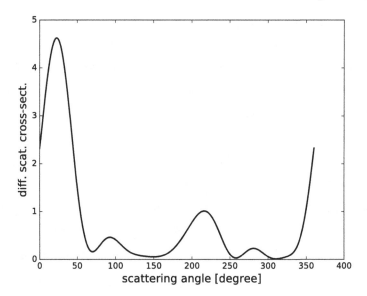

Fig. 3.51 Differential scattering cross-sections of a rotated s-p Janus sphere. Parameters: radius $a = 1.0$ mm, size parameter $\beta = 3.0$, splitting angle $\theta_J = 90.0°$, Eulerian angles of rotation ($\alpha = 0°$, $\theta_p = 90°$), truncation parameters $ncut = 26$, $lcut = 4$. Total scattering cross-section: $\sigma_{tot} = 6.34$

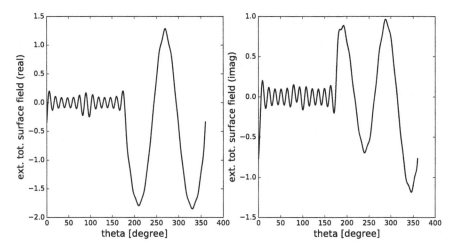

Fig. 3.52 Test of the homogeneous Dirichlet condition of the Janus sphere of Fig. 3.51. Truncation parameter: $ncut = 26$

Table 3.2 Test of the reciprocity condition for the Janus sphere of Fig. 3.51. $\alpha = 0°$ was used in both configurations

Config.	$\theta_p = 0°\ \theta = 90°$	$\theta_p = 90°\ \theta = 270°$
$ncut = 5\ (nonconvergent)$	0.152	0.152
$ncut = 26\ (convergent)$	0.149	0.149

intensity at a scattering angle of $\theta = 23°$ that goes again beyond the intensity of the forward direction. This would produce also a halo effect if looking from above on an ensemble of independently scattering s-p Janus spheres, and if these spheres are randomly oriented with respect to the Eulerian angle α but at a fixed $\theta_p = 90°$. Regarding the fulfillment of the reciprocity condition we have to state the same behavior as discussed in Sect. 3.1.1. This can be seen from Table 3.2 where one can find the differential scattering cross sections of two reciprocal configurations for the Janus sphere of Fig. 3.51, and if the computations are performed with two different truncation parameters that result in a nonconvergent and a convergent result.

3.3 Janus Sphere of h-p Type

3.3.1 Axisymmetric Orientation

Now, transmission condition (2.26) must be used for the h-p Janus sphere to express the expansion coefficients g_n in terms of the coefficients c_n and d_n. This results from the fact that the Neumann condition must hold on its whole surface. Instead of (3.26) we thus get

$$g_n = \frac{\kappa}{j_n'(\beta_p)} \cdot \left\{ c_n \cdot \left[h_n^{(1)}(\beta) \right]' + d_n \cdot j_n'(\beta) \right\}. \tag{3.34}$$

This expression, along with the corresponding field expansions and the remaining boundary conditions (2.21) and (2.25), is then condensed into the generalized boundary condition (3.1). This gives

$$\sum_{n=0}^{ncut} c_n \cdot \left\{ \left[h_n^{(1)}(\beta) \right]' \cdot f_<(\theta) + \left\{ h_n^{(1)}(\beta) - \frac{\kappa \cdot \left[h_n^{(1)}(\beta) \right]' \cdot j_n(\beta_p)}{j_n'(\beta_p)} \right\} \cdot f_>(\theta) \right\} \cdot$$

$$Y_{0n}(\theta, \phi) = -\sum_{n=0}^{ncut} d_n \cdot \left\{ j_n'(\beta) \cdot f_<(\theta) + \left[j_n(\beta) - \frac{\kappa \cdot j_n'(\beta) \cdot j_n(\beta_p)}{j_n'(\beta_p)} \right] \cdot \right.$$

$$\left. f_>(\theta) \right\} \cdot Y_{0n}(\theta, \phi) \tag{3.35}$$

for the Janus sphere of h-p type if axisymmetrically oriented in the laboratory frame. The corresponding T-matrix reads therefore

$$\mathbf{T_{hp}} = -\left[\mathbf{Y_<} \cdot \mathbf{D_{h'}} + \mathbf{Y_>} \cdot \mathbf{D_c}\right]^{-1} \cdot \left[\mathbf{Y_<} \cdot \mathbf{D_{j'}} + \mathbf{Y_>} \cdot \mathbf{D_d}\right] , \tag{3.36}$$

where the two matrices $\mathbf{D_c}$ and $\mathbf{D_d}$ are given by

$$\mathbf{D_c} : h_n^{(1)}(\beta) - \frac{\kappa \cdot \left[h_n^{(1)}(\beta)\right]' \cdot j_n(\beta_p)}{j_n'(\beta_p)} \quad n = 0, \ldots, ncut \tag{3.37}$$

$$\mathbf{D_d} : j_n(\beta) - \frac{\kappa \cdot j_n'(\beta) \cdot j_n(\beta_p)}{j_n'(\beta_p)} \quad n = 0, \ldots, ncut . \tag{3.38}$$

The expansion coefficients of the scattered field are computed from

$$\vec{c}^{\,tp} = \mathbf{T_{hp}} \cdot \vec{d}^{\,tp} . \tag{3.39}$$

The T-matrices of the homogeneous sound hard and sound penetrable sphere result from expression (3.36) if choosing $\theta_J = 180°$ or $\theta_J = 0°$, respectively.

Regarding its convergence behavior, this type of Janus sphere shows an effect which also appeared at the other types of Janus spheres, but which is most clearly expressed at this type. It is of some worth to take a closer look on it, because this effect once again makes it clear that one has to spend a much higher effort to produce a convergent result for Janus spheres than it is required for the spherical scatterers considered in the second chapter. This effect is expressed in Figs. 3.53, 3.54 and 3.55. It can also be observed in Figs. 3.4/3.12 of the h-s Janus sphere and in Figs. 3.30/3.37

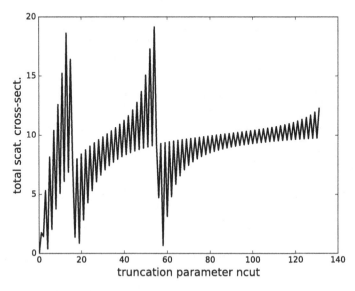

Fig. 3.53 Total scattering cross-section of a h-p Janus sphere in axisymmetric orientation as a function of the truncation parameter *ncut*. Parameters: radius $a = 1.0$ mm, size parameter $\beta = 1.0$, $\kappa_k = \rho_p = 1.5$, splitting angle $\theta_J = 90.0°$

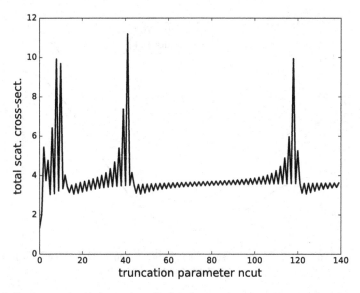

Fig. 3.54 Total scattering cross-section of a h-p Janus sphere in axisymmetric orientation as a function of the truncation parameter $ncut$. Parameters: radius $a = 1.0\,$mm, size parameter $\beta = 3.0$, $\kappa_k = \rho_p = 1.5$, splitting angle $\theta_J = 90.0°$

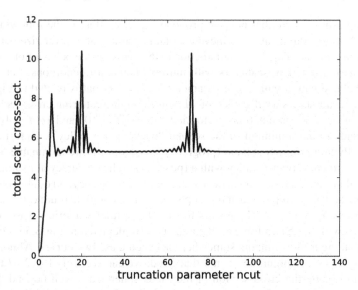

Fig. 3.55 Total scattering cross-section of a h-p Janus sphere in axisymmetric orientation as a function of the truncation parameter $ncut$. Parameters: radius $a = 1.0\,$mm, size parameter $\beta = 6.0$, $\kappa_k = \rho_p = 1.5$, splitting angle $\theta_J = 90.0°$

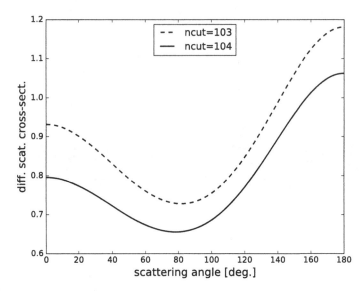

Fig. 3.56 Differential scattering cross-sections of the Janus sphere of Fig. 3.53 for two successive truncation parameters

of the s-p Janus sphere, but not that pronounced. The effect is the following: It is obviously more complicate to achieve a certain accuracy at a lower size parameter than at a higher one. This is in contradiction to the convergence behavior of spherical as well as nonspherical scatterers with uniform boundary conditions. Let us first consider the situation with a size parameter of $\beta = 1.0$ that is related to Fig. 3.53. The best result shows a difference of 11% in the total scattering cross-section for the two successive computations performed with $ncut = 103$ and $ncut = 104$ (this corresponds to the minimum in the second "convergence valley"). Unfortunately, with the Python program at hand (program *janus_axial.py*) we are unable to cover also the third convergence valley with a (possibly) higher accuracy at its minimum. This would require a more sophisticated program that goes beyond double precision accuracy, and that employs a more sophisticated integration routine to calculate the matrices $\mathbf{Y}_<$ and $\mathbf{Y}_>$. The results for the differential scattering cross-sections calculated with these two truncation parameters are depicted in Fig. 3.56. One may expect that the truth is running somewhere between these two curves. Although the accuracy for this configuration is not that good, we can see from Fig. 3.57 that the differential scattering cross-section of the Janus sphere runs well beyond the two limiting cases. That we do not move completely beside the track with the results of Fig. 3.56 shows the good fulfillment of the homogeneous Neumann and Dirichlet condition on the respective sound hard and sound penetrable part of the surface, as it can be seen from Figs. 3.58 and 3.59. Regarding the fulfillment of the homogeneous Neumann condition that belongs to the penetrable part of the surface we have to state

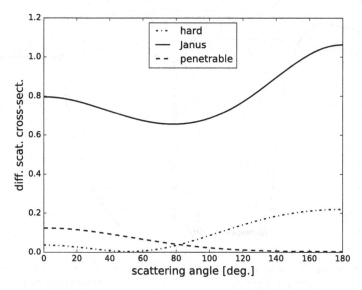

Fig. 3.57 Differential scattering cross-sections of the Janus sphere of Fig. 3.53 compared to the scattering behavior of the two limiting cases of the sound hard and penetrable sphere. *ncut* = 104 was used for the Janus sphere

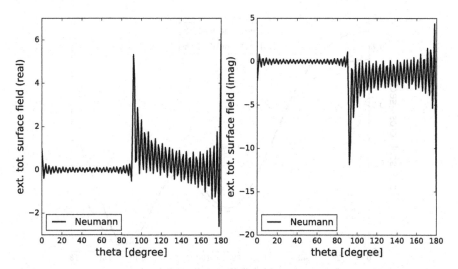

Fig. 3.58 Test of the homogeneous Neumann condition of the Janus sphere of Fig. 3.57

again an accuracy within double precision similar to that one depicted in Fig. 3.34. This is a consequence of relation (3.34), of course.

The corresponding results for this Janus sphere but at a size parameter of $\beta = 3.0$ are given in Figs. 3.60, 3.61, 3.62, 3.63 and 3.64. Now we achieved already a difference of only 3.3% in the total scattering cross-section for the two successive com-

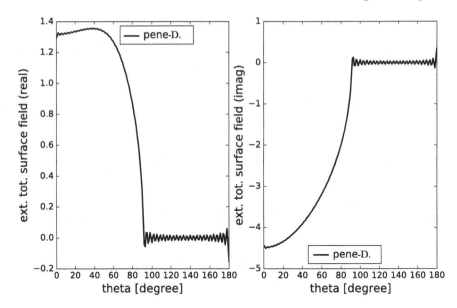

Fig. 3.59 Test of the homogeneous Dirichlet condition that belongs to the sound penetrable part of the Janus sphere of Fig. 3.57

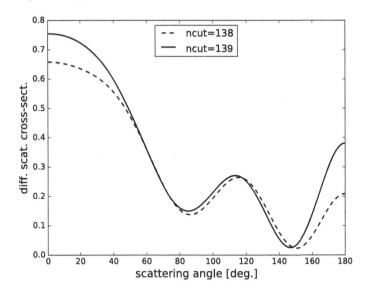

Fig. 3.60 Differential scattering cross-sections of the Janus sphere of Fig. 3.54 for two successive truncation parameters. Total scattering cross-sections: $\sigma_{tot} = 3.397$ ($ncut = 138$), $\sigma_{tot} = 3.613$ ($ncut = 139$)

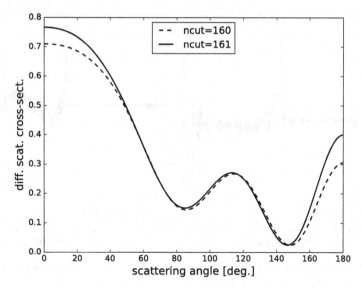

Fig. 3.61 Differential scattering cross-sections of the Janus sphere of Fig. 3.54 for two successive truncation parameters. Total scattering cross-sections: $\sigma_{tot} = 3.52$ ($ncut = 160$), $\sigma_{tot} = 3.637$ ($ncut = 161$)

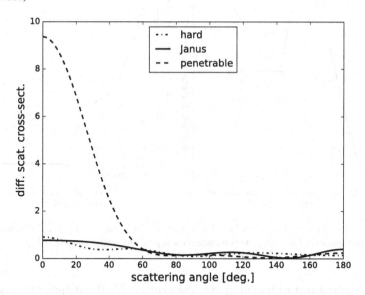

Fig. 3.62 Differential scattering cross-sections of the Janus sphere of Fig. 3.54 compared to the scattering behavior of the two limiting cases of the sound hard and penetrable sphere. $ncut = 161$ was used for the Janus sphere

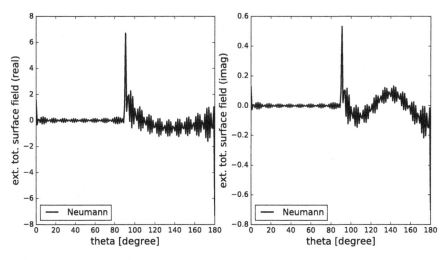

Fig. 3.63 Test of the homogeneous Neumann condition of the Janus sphere of Fig. 3.62. The computation was performed with *ncut* = 161

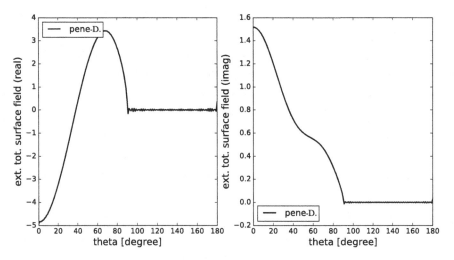

Fig. 3.64 Test of the homogeneous Dirichlet condition that belongs to the sound penetrable part of the Janus sphere of Fig. 3.62. The computation was performed with *ncut* = 161

putations performed with *ncut* = 160 and *ncut* = 161. But despite the very large truncation parameters used in these computations there still remain larger differences in the forward- and backscattering region in the differential scattering cross-sections. Comparing Figs. 3.60 and 3.61, we find that the differential scattering cross-section converges only very slowly. To improve the accuracy even further requires again a more sophisticated program that can run higher truncation parameters. On the other hand, Figs. 3.63 and 3.64 demonstrate the amazingly good fulfillment of the homoge-

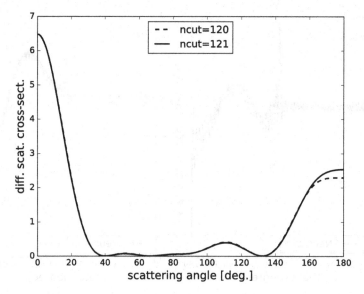

Fig. 3.65 Differential scattering cross-sections of the Janus sphere of Fig. 3.55 for two successive truncation parameters. Total scattering cross-sections: $\sigma_{tot} = 5.32$ ($ncut = 120$), $\sigma_{tot} = 5.307$ ($ncut = 121$)

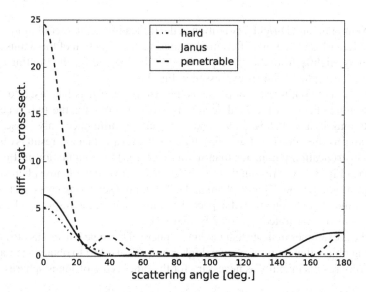

Fig. 3.66 Differential scattering cross-sections of the Janus sphere of Fig. 3.55 compared to the scattering behavior of the two limiting cases of the sound hard and penetrable sphere. $ncut = 121$ was used for the Janus sphere

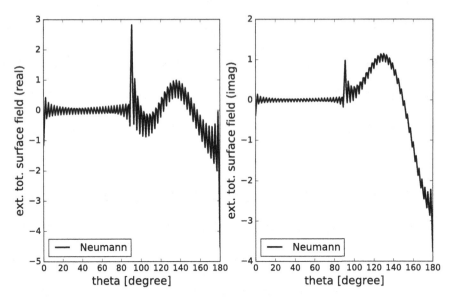

Fig. 3.67 Test of the homogeneous Neumann condition of the Janus sphere of Fig. 3.66. The computation was performed with $ncut = 121$

neous Neumann- and Dirichlet condition at the truncation parameter of $ncut = 161$. It is also remarkable that the differential scattering cross-section of this Janus sphere differs only slightly from the corresponding result obtained for the limiting case of the sound hard sphere. This is expressed in Fig. 3.62.

The situation continues to improve as the size parameter is increased to $\beta = 6$. The difference between the total scattering cross-sections for the two successive computations is now below 0.3%, and it remains a difference only close to the backscattering direction (see Fig. 3.65). It is also interesting that this result is obtained at a lower truncation parameter than it was used for the computation in Fig. 3.60. Looking at Fig. 3.66 it turns out that the differences between the Janus sphere and the limiting case of the sound hard sphere in the differential scattering cross-sections are again not that large as one would expect. The fulfillment of the respective boundary conditions is demonstrated in Figs. 3.67 and 3.68.

All these examples indicate that the Janus sphere of h-p type is a not easy to analyze scattering object that requires a much higher numerical effort to obtain accurate and trustworthy results than it was necessary for the other types of Janus spheres.

3.3.2 Arbitrary Rotation

Since the T-matrix of the h-p Janus sphere in the rotated system can be derived in the same way as described in Sect. 3.2.2 we just have to add the superscript (l) to

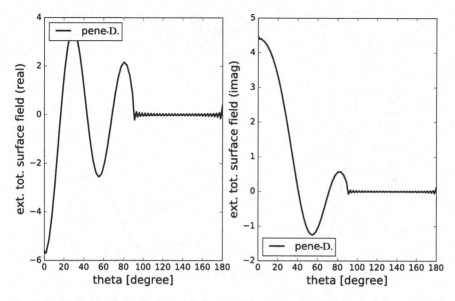

Fig. 3.68 Test of the homogeneous Dirichlet condition that belongs to the sound penetrable part of the Janus sphere of Fig. 3.66. The computation was performed with $ncut = 121$

expression (3.36). That is, the T-matrix of the arbitrarily rotated h-p Janus sphere reads

$$\mathbf{T}_{\mathbf{hp}}^{(l)} = - \left[\mathbf{Y}_<^{(l)} \cdot \mathbf{D}_{h'} + \mathbf{Y}_>^{(l)} \cdot \mathbf{D}_c \right]^{-1} \cdot \left[\mathbf{Y}_<^{(l)} \cdot \mathbf{D}_{j'} + \mathbf{Y}_>^{(l)} \cdot \mathbf{D}_d \right] , \qquad (3.40)$$

and the expansion coefficients of the scattered field are calculated according to (in matrix notation)

$$\vec{c}_{\bar{l}}^{\,tp} = \mathbf{T}_{\mathbf{hp}}^{(\bar{l})} \cdot \vec{d}_{\bar{l}}^{\,tp} ; \quad \bar{l} = -lcut, \ldots, lcut \qquad (3.41)$$

in the rotated system.

Now it is even more complicate to calculate the two matrices $\mathbf{Y}_<^{(l)}$ and $\mathbf{Y}_>^{(l)}$ according to (3.22) and (3.23) since larger values of the azimuthal mode l and the truncation parameter $ncut$ must be considered especially at lower size parameters. This requires once again more sophisticated routines for the special functions and the integration than used in the programs at hand. Using these programs a sufficiently accurate result is therefore only obtained for the h-p Janus sphere of Fig. 3.55 if the truncation parameter $ncut = 37$ is used. This value was chosen according to the discussion in Sect. 3.1.1 regarding the possibility to lower the truncation parameter for a given scattering configuration of Janus spheres. Figure 3.69 shows the good agreement of the differential scattering cross-sections obtained with 2 different truncation parameters. Only in the backscattering region we have a smaller deviation between the two results. The determination of the truncation parameter $lcut$ for this Janus sphere if rotated by the Eulerian angles ($\alpha = 0°, \theta_p = 90°$) provides $lcut = 6$. The differen-

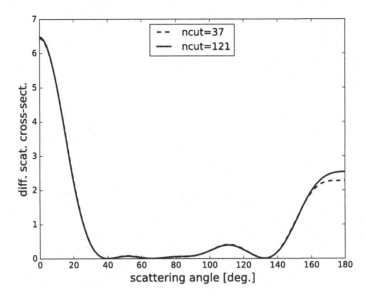

Fig. 3.69 Differential scattering cross-sections of the Janus sphere of Fig. 3.55 in axisymmetric orientation. The computation was performed with the two truncation parameters $ncut = 37, 121$

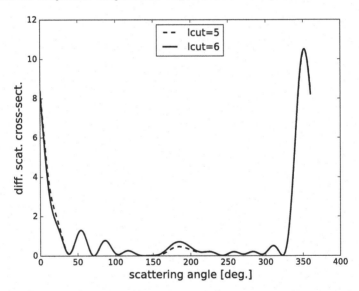

Fig. 3.70 Differential scattering cross-sections of the rotated Janus sphere of Fig. 3.55. Eulerian angles: ($\alpha = 0°, \theta_p = 90°$). The computation was performed with $ncut = 37$ and the two truncation parameters $lcut = 5, 6$ of the azimuthal modes. Total scattering cross-sections: $\sigma_{tot} = 5.631$ ($lcut = 5$), $\sigma_{tot} = 5.756$ ($lcut = 6$)

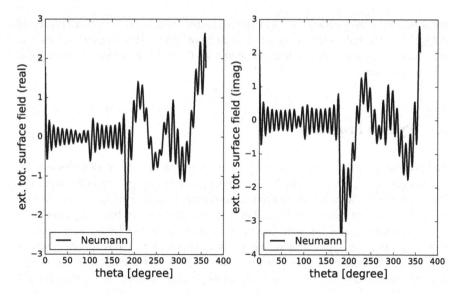

Fig. 3.71 Approximation of the Neumann boundary condition of the Janus sphere of Fig. 3.70. Truncation parameter of the azimuthal modes: $lcut = 6$

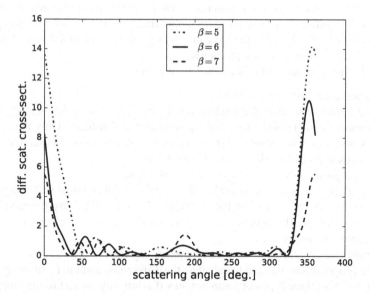

Fig. 3.72 Differential scattering cross-sections of the rotated Janus sphere of Fig. 3.70. $ncut = 37$ and $lcut = 6$ was used in these computations. It shows the evolution of the halo effect for 3 different size parameters

tial scattering cross-sections for the two successive values $lcut = 5, 6$ are presented in Fig. 3.70. The halo phenomenon can again be observed. How does it evolves with increasing size parameters is demonstrated with Fig. 3.72. It is most pronounced at the size parameter of $\beta = 6$ (Fig. 3.71).

3.4 Python Programs

The programs of this chapter are organized in the same way as described at the beginning of Sect. 2.5. First one has to run the program that generates the input file. The programs to calculate the differential and total scattering cross-sections as well as the program that can be used to test the approximation of the boundary conditions of the Janus sphere of h-s type can be used afterwards. But I have to emphasize once again that the application of the T-matrix approach to Janus spheres requires a much higher numerical effort (and also more sophisticated numerical procedures than provided with the programs at hand) to generate physically meaningful and numerically trustworthy results over larger parameter regions. Of course, one can try to apply a different method like boundary integral equation techniques to analyze the scattering behavior of Janus spheres. Such an approach is presented in [10], for example. But, unfortunately, no scattering data are given and discussed in this paper. However, it is not only my experience that, using a different method will often only transform rather than solve the problems.

The Python programs of this chapter are the following:

- programs *janus_spheres_ncut.py*
  This program calculates the total scattering cross-section in dependence on the truncation parameter *ncut* for all three types of Janus spheres. Only the axisymmetric orientation is considered in this program since it is aimed at supporting the convergence procedure described in Sect. 3.1.1.
- programs *janus_axial.py* and *janus_axial_input.py*:
  These programs can be used to calculate the differential and total scattering cross-sections of all three types of Janus spheres if axisymmetrically oriented in the laboratory frame. The computation is performed at a given size parameter β and truncation parameter *ncut*.
- programs *janus_rotated.py* and *janus_rotated_input.py*:
  These programs can be used to calculate the differential and total scattering cross-sections of all three types of Janus spheres if arbitrarily rotated in the laboratory frame. The computation is performed at a given size parameter β. The two truncation parameters *ncut* and *lcut* must also be given by the user. Of course, these programs can also be applied to Janus spheres in axisymmetric orientation if choosing the Eulerian angles ($\alpha = 0°$, $\theta_p = 0°$) and $lcut = 0$. However, I found it appropriate to have two independent programs to test the fulfillment of the reciprocity condition.

- programs *h_s_axial_surface.py*:
  This program calculates the surface fields of the h-s Janus sphere to test the fulfill-ment of the corresponding boundary conditions on each subsurface (homogeneous Neumann and Dirichlet condition). Extending this program to become applicable also to the other types of Janus spheres is left to the reader as an exercise. In so doing, please use expressions (3.26) and (3.34), respectively, for the computation of the expansion coefficients of the internal field that must additionally be taken into account in the case of the s-p- or the h-p Janus sphere.

Several examples of how to use these programs have been already presented in this chapter. The full programs are given in Appendix C.

References

1. Sommerfeld, A.: Partial Differential Equations in Physics. Academic Press, New York (1949)
2. Wauer, J.: A least square T-matrix method for non-spherical dielectric particle compared with Waterman's T-matrix method in the near and far field. JQSRT **180**, 47–54 (2016)
3. Rother, T., Kahnert, M.: Electromagnetic Wave Scattering on Nonspherical Particles: Basic Methodology and Applications. Springer, Heidelberg (2014)
4. Abramowitz, M., Stegun, I.A.: Handbook of Mathematical Functions. Harri Deutsch, Frank-furt/Main (1984)
5. Zhang, J., Grzybowski, B.A., Granick, S.: Janus particle synthesis, assembly, and application. Langmuir **33**, 6964–6977 (2017)
6. Kalhor, H.A.: Electromagnetic scattering by a dielectric slab loaded with a periodic array of strips over a ground plane. IEEE Trans. Antennas Propag. **36**(1), 147–151 (1988)
7. Pregla, R., Pascher, W.: The method of lines. In: Itoh, T. (ed.) Numerical Techniques for Microwave and Millimeter-Wave Passive Structures. Wiley, New York (1989)
8. Kristensson, G., Waterman, P.C.: The T-matrix for acoustic and electromagnetic scattering by circular disks. J. Acoust. Soc. Am. **72**, 1612–1625 (1982)
9. Baars, W.M.: Diffraction of sound waves. Acustica **14**, 289–296 (1964)
10. Gillman, A.: An integral equation technique for scattering problems with mixed boundary conditions. Adv. Comput. Math. **43**, 351–364 (2017)

Chapter 4
Scattering on Bispheres

Plane wave scattering on two homogeneous spheres is a first step beyond the well-known and well-understood scattering problem on a single homogeneous sphere that was discussed in the second chapter of this book. It allows us to introduce the concept of modeling multiple scattering processes in the context of the T-matrix method. A first solution of this problem for two sound soft spheres in an axisymmetric orientation was already given in [1], but without using the addition theorem for the eigenfunctions of the underlying Helmholtz equation. Solving this scattering problem by using a multipole expansion of the involved fields and the addition theorem was described in detail later on in [2]. This approach was applied in [3, 4], for example, to analyze the scattering behavior of sound soft and sound hard bisphere combinations in different but fixed orientations. A detailed list of references regarding this scattering problem and the different ways of solving it can be found in [5]. However, there is to state a lack of systematic studies and of an elaborate data base which, in contrast, exist for the corresponding scattering problem of electromagnetic waves on bispheres (see [6], for example, and the cited references therein). It is one goal of this chapter to improve this situation by providing solutions for different combinations of sound soft, sound hard, and sound penetrable spheres in a bisphere configuration, and to discover bispheres in fixed orientations as well as ensembles of bispheres.

In their analysis of light scattering on randomly oriented bispheres Mishchenko and Mackowski found that the averaged scattering cross-section of bispheres is nearly identical with the independent scattering of the component spheres [6]. And since the rigorous solution of the scattering problem on bispheres still requires a considerable numerical effort, especially if used in applications of practical interest, it is another goal of this chapter to identify scattering configurations where we can benefit from simplifications. This is achieved by applying an iteration scheme to the derived T-matrix equations. The importance and accuracy of the iterative solutions are dis-

Electronic supplementary material The online version of this chapter
(https://doi.org/10.1007/978-3-030-36448-9_4) contains supplementary material, which is available to authorized users.

© Springer Nature Switzerland AG 2020
T. Rother, *Sound Scattering on Spherical Objects*,
https://doi.org/10.1007/978-3-030-36448-9_4

cussed, and scattering configurations are identified which can be treated within an appropriate accuracy by a certain order of iteration. In so doing, it turns out that there are some configurations of practical interest which benefit already from the zero-order iteration. It is shown that this zero-order iteration can simply be expressed by use of the T-matrices of the homogeneous component spheres of a certain bisphere configuration, together with an analytical phase term that considers the orientation of the component spheres in the laboratory frame. It is an essential advantage of this expression that it is, in principle, not restricted to treat only bispheres but any number of spheres, nonspherical scatterers, and even Janus spheres. The latter situation requires the knowledge of the T-matrices of the single homogeneous but nonspherical scatterers, of course (see Sect. 2.4).

In Sect. 1.4 of this book we have already mentioned the fact that any local shift of an obstacle appears only as a phase term in the far-field of the laboratory frame, and that this shift is washed out if calculating the differential scattering cross-sections. Regarding arbitrarily oriented bispheres we can therefore benefit from a further simplification throughout this chapter: It is assumed without loss of generality that one of the spheres of any bisphere configuration is always centered in the laboratory frame. And, moreover, if the considered bisphere consists of two spheres with different radii this will always be the sphere with the larger radius.

4.1 A Simple Approximation

It is shown in [7] that the scattering amplitude function of the famous double-slit experiment in Fraunhofer approximation can be described by the sum of the scattering amplitudes of the single slits multiplied by an additional phase term that describes the positions of the slits in the slit plane. The edge effects and the interaction between the slits are neglected in this approximation. However, it is well-known that experimental results in the far-field and near the forward direction agree quite well with this simple approximation. One may therefore ask if there exists a similar approximation for bispheres that neglects the interaction between the spheres, but that takes the morphology of the single spheres together with the geometrical configuration of the bisphere and the scattering angle in the far-field of the laboratory frame into account. That such a possibility exists in fact will now be shown. Due to the above mentioned simplification it is only necessary to link this geometrical phase term to the non-centered sphere. Let us therefore assume that

$$f_1(\theta) = e^{ik_0\vec{b}\cdot(\hat{k}_i - \hat{r}_\theta)} \cdot \widehat{f_1}(\theta) \tag{4.1}$$

holds for the scattering amplitude function $f_1(\theta)$ of the shifted sphere in the laboratory frame, while $\exp(ik_0\vec{b}\cdot\hat{k}_i)\cdot\widehat{f_1}(\theta)$ is the scattering amplitude of this sphere in its local frame (i.e., the frame this sphere is centered in).

$$\vec{b} = b\cdot\left(\cos\theta_p\cdot\hat{z} + \sin\theta_p\cdot\cos\alpha\cdot\hat{x} + \sin\theta_p\cdot\sin\alpha\cdot\hat{y}\right) \tag{4.2}$$

Fig. 4.1 Bisphere
configuration in the
x-z-plane of the laboratory
frame. $(\alpha = 0, \theta_p)$ are the
Eulerian angles of
orientation. b corresponds to
the center distance of both
spheres, and θ denotes the
scattering angle

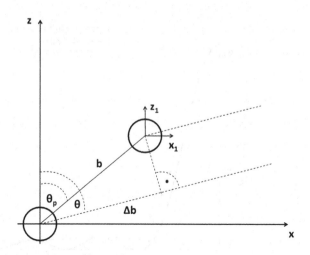

represents the vector from the origin of the laboratory frame to the center of the
shifted sphere.

$$\hat{k}_i = \hat{z} \tag{4.3}$$

is the directional unit vector of the incident plane wave in the laboratory frame, and

$$\hat{r}_\theta = \cos\theta \cdot \hat{z} + \sin\theta \cdot \hat{x} \tag{4.4}$$

is the directional unit vector in the scattering plane (the x-z-plane) of the laboratory
frame to a certain point in the far-field. The situation for a bisphere configuration in
the scattering plane of the laboratory frame is shown in Fig. 4.1. Note, that Δb is the
scalar product of

$$\Delta b = \vec{b} \cdot \hat{r}_\theta \,, \tag{4.5}$$

and that the scattering angles θ are identical in both local systems. The term $\exp(ik_0\vec{b} \cdot \hat{k}_i)$ in (4.1) is simply a consequence of the phase shift of the primary incident plane
wave that must be taken into account in the local frame of the shifted sphere (see
(1.81) in Sect. 1.3.1). The total scattering amplitude function of a certain bisphere
configuration in this approximation is simply the sum

$$f_t(\theta) = f_0(\theta) + f_1(\theta) \,, \tag{4.6}$$

where $f_0(\theta)$ is the scattering amplitude function of the sphere that is centered in
the laboratory frame. Both the local scattering amplitude functions $f_0(\theta)$ and $\widehat{f}_1(\theta)$
are calculated according to (1.84) with $d_{l'n'}$ given by (2.6), and with the T-matrices
given by (2.5), (2.22), and (2.28). Which of these T-matrices is used depends on the
morphology of the considered component sphere, of course.

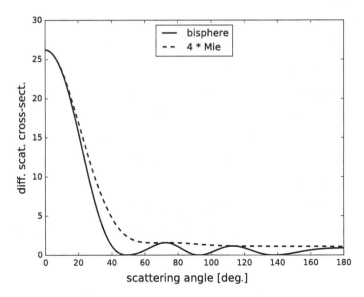

Fig. 4.2 Differential scat. cross-sect. of a (s,s)-bisphere. Parameters: $a_0 = a_1 = 1.0$ mm, $\beta_0 = \beta_1 = 3.0$, $b = 3.0$ mm, $\alpha = 0°$, $\theta_p = 0°$. Total scattering cross-section: $\sigma_{tot} = 17.625$, differential scattering cross-section at $\theta = 90°$: $\partial\sigma/\partial\Omega = 0.0638$

This approximation is implemented in the Python program *bisphere_approxy.py*. Let us apply it to four different bisphere configurations. We will denote any combination of bispheres with "(a,b)". "a = s, h, or p" characterizes the sphere which is centered in the laboratory frame, and "b = s,h, or p" characterizes the shifted sphere. The differential and total scattering cross-sections of two identical sound soft spheres (a (s,s)-bisphere) with the radii $a_0 = a_1 = 1.0$ mm, with size parameters of $\beta_0 = \beta_1 = 3.0$, and with a center distance of $b = 3.0$ mm between the spheres are presented in Fig. 4.2. It was moreover assumed that the bisphere is axisymmetrically oriented (i.e., the shifted sphere is located on the z-axis of the laboratory frame). Figure 4.3 shows the corresponding results for the same combination but if the shifted sphere is now located on the x-axis of the laboratory frame. However, the z-axis of the laboratory frame is still the axis of symmetry with respect to the differential scattering cross-section for both these orientations. It was therefore sufficient to plot the results in the interval $\theta \in [0°, 180°]$. The fourfold of the Mie result of the single sound soft sphere is also plotted for intercomparison purposes with the known behavior of two identical double slits. From this latter combination we know that the fourfold intensity of the single slit forms the envelope of the double slit (see [7], Chap. 4.2 therein, for example). The same obviously applies to the two identical sound soft spheres. According to the reciprocity condition, the differential scattering cross-section at a scattering angle of $\theta = 90°$ for the bispheres in axisymmetric orientation must be identical with the differential scattering cross-section at the scattering angle of $\theta = 90°$ for the bisphere configuration of Fig. 4.3. That this is indeed

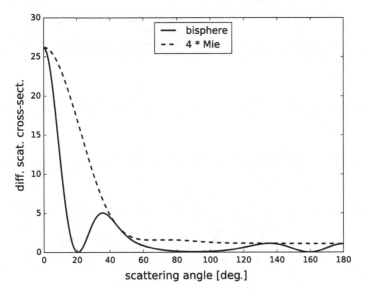

Fig. 4.3 Differential scat. cross-sect. of a (s,s)-bisphere. Parameters: $a_0 = a_1 = 1.0$ mm, $\beta_0 = \beta_1 = 3.0, b = 3.0$ mm, $\alpha = 0°, \theta_p = 90°$. Total scattering cross-section: $\sigma_{tot} = 17.625$, differential scattering cross-section at $\theta = 90°$: $\partial\sigma/\partial\Omega = 0.0638$

the case can be seen from the corresponding numbers given in the figure captions. That is, the reciprocity condition holds already for this simple approximation! We can observe moreover that the total scattering cross-section is independent of the orientation of the bispheres.

Let us now see what happens if we replace the shifted sphere in the above example by a sound hard sphere, i.e., if we consider the (s,h)-bisphere combination. The results are shown in Figs. 4.4 and 4.5. Regarding the orientation with the shifted sphere located on the x-axis of the laboratory frame, now the z-axis is no longer the axis of symmetry with respect to the differential scattering cross-section. This can clearly be seen from Fig. 4.5. The reciprocity condition holds also for this bisphere combination. But now we have to compare the differential scattering cross-section at a scattering angle of $\theta = 90°$ of the axisymmetric orientation with the differential scattering cross-section at the scattering angle of $\theta = 270°$ for the bisphere configuration of Fig. 4.5. This is a consequence of the asymmetric behavior of the latter configuration. That twice the result of the single sound soft sphere plus twice the result of the single sound hard sphere forms the envelope of this bisphere configuration gets lost, on the other hand. And, finally, we can again observe a halo phenomenon for the configuration of Fig. 4.5 close to the forward direction, as it was already observed for the Janus spheres.

The question about the importance of this simple approximation will be answered in the next section, where its results are compared with the respective results of a more rigorous approach. But it should already be emphasized at this point that this

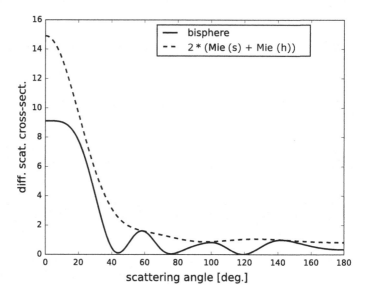

Fig. 4.4 Differential scat. cross-sect. of a (s,h)-bisphere. Parameters: $a_0 = a_1 = 1.0$ mm, $\beta_0 = \beta_1 = 3.0$, $b = 3.0$ mm, $\alpha = 0°$, $\theta_p = 0°$. Total scattering cross-section: $\sigma_{\text{tot}} = 12.06$, differential scattering cross-section at $\theta = 90°$: $\partial\sigma/\partial\Omega = 0.6034$

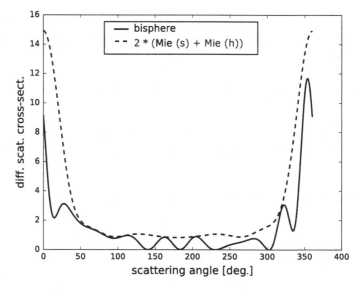

Fig. 4.5 Differential scat. cross-sect. of a (s,h)-bisphere. Parameters: $a_0 = a_1 = 1.0$ mm, $\beta_0 = \beta_1 = 3.0$, $b = 3.0$ mm, $\alpha = 0°$, $\theta_p = 90°$. Total scattering cross-section: $\sigma_{\text{tot}} = 12.06$, differential scattering cross-section at $\theta = 270°$: $\partial\sigma/\partial\Omega = 0.6034$

approximation does not requires the application of the separation matrix and the matrix of rotation! The T-matrix of each single sphere in its local system is the only quantity that is needed. And this approximation can simply be generalized to any number of spheres and also to nonspherical objects if the corresponding T-matrices are known.

4.2 T-Matrix Equations and Iterative Solutions

We will now derive the T-matrix equations that can iteratively be solved for any bisphere combination. Here we want to treat the two cases of the axisymmetrically and arbitrarily oriented bispheres separately from each other. But the latter case contains the axisymmetric orientation as a limiting case, of course. This splitting is due to the fact that the axisymmetric orientation benefits from simplifications that result in a lower numerical effort. It is moreover of some worth to have an independent approach for both these situations since the more simple situation of the axisymmetric orientation will allow us to test the correct numerical implementation of the more complex situation of arbitrarily oriented bispheres.

4.2.1 Axisymmetric Orientation

Let us start with the simpler situation of an axisymmetric bisphere configuration, as depicted in Fig. 4.6. Only the boundary conditions for two sound soft spheres are considered in the following derivation. However, we can walk along the same way for all the other boundary conditions. The finally derived T-matrix equation holds therefore for all the different combinations of sound soft, sound hard, and sound penetrable spheres since expressed in terms of the T-matrices of each sphere in its local system, as given in Sect. 2.1.

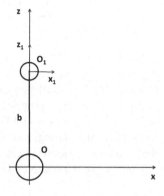

Fig. 4.6 Two spheres with different radii a_0 and a_1 located on the z-axis of the laboratory frame **O** with a center distance b. $\mathbf{O_1}$ denotes the local frame of the shifted sphere with radius a_1

$$u_{s_0}(k_0 r, \theta) = \sum_{m_0=0}^{ncut} c_{m_0} \cdot \varphi_{0\,m_0}(k_0 r, \theta) \tag{4.7}$$

and

$$u_{s_1}(k_0 r_1, \theta_1) = \sum_{m_1=0}^{ncut} q_{m_1} \cdot \varphi_{0\,m_1}(k_0 r_1, \theta_1) \tag{4.8}$$

are the expansions of the scattered fields in the local system of each sphere (these are the laboratory frame **O** and the shifted frame **O**$_1$ in Fig. 4.6) produced by the incident plane wave (1.68). Note again that only the azimuthal mode $l = 0$ must be considered for the axisymmetric orientation! These scattered fields represent the additionally existing incident fields in the local system of the respective other sphere. Applying the boundary condition (2.1) for a sound soft sphere in each of the local systems, and if taking the transformation relation (1.27) for a shift along the z-axis into account result in the following two equations:

- In the laboratory frame **O** at $r = a_0$:

$$\sum_{m_0=0}^{ncut} c_{m_0} \cdot \varphi_{0\,m_0}(\beta_0, \theta) = - \sum_{m_0=0}^{ncut} d_{m_0} \cdot \psi_{0\,m_0}(\beta_0, \theta) -$$
$$\sum_{m_1=0}^{ncut} \sum_{m_1'=0}^{ncut} q_{m_1} \cdot (-1)^{m_1'} \cdot S^0_{m_1 m_1'}(b) \cdot \psi_{0\,m_1'}(\beta_0, \theta) .$$

$$\tag{4.9}$$

- In the local system **O**$_1$ of the shifted sphere at $r = a_1$:

$$\sum_{m_1=0}^{ncut} q_{m_1} \cdot \varphi_{0\,m_1}(\beta_1, \theta_1) = - \sum_{m_1=0}^{ncut} \hat{d}_{m_1} \cdot \psi_{m_1}(\beta_1, \theta_1) -$$
$$\sum_{m_0=0}^{ncut} \sum_{m_0'=0}^{ncut} c_{m_0} \cdot (-1)^{m_0} \cdot S^0_{m_0 m_0'}(b) \cdot \psi_{0\,m_0'}(\beta_1, \theta_1) .$$

$$\tag{4.10}$$

d_{m_0} and $\hat{d}_{m_1}$ are the expansion coefficients of the incident plane wave according to (1.70) and (1.79). Similar equations are obtained from the other boundary conditions. These equations are multiplied in the next step by the spherical harmonics $Y_{0\,\tilde{m}_0}(\theta, \phi)$ or $Y_{0\,\tilde{m}_0}(\theta_1, \phi_1)$ given by (1.13) and integrated subsequently over the respective spherical surface. In so doing, and due to the orthonormality relation (1.14) we end up with the following equation system to determine the so far unknown expansion coefficients c_{m_0} and q_{m_1} of the scattered fields:

$$c_{m_0} - [T_O(\beta_0)]_{m_0 m_0} \cdot (-1)^{m_0} \sum_{m_1=0}^{ncut} q_{m_1} \cdot S^0_{m_0 m_1}(b) = c^{(0)}_{m_0} \; ; \; m_0 = 0, \ldots, ncut$$

$$\tag{4.11}$$

$$q_{m_1} - \left[T_{O_1}(\beta_1)\right]_{m_1 m_1} \cdot \sum_{m_0=0}^{ncut} c_{m_0} \cdot (-1)^{m_0} \cdot S^0_{m_0 m_1}(b) = q^{(0)}_{m_1} \; ; \; m_1 = 0, \ldots, ncut$$

$$\tag{4.12}$$

where

$$c^{(0)}_{m_0} = \sum_{\tilde{m}_0=0}^{ncut} [T_O(\beta_0)]_{m_0 \tilde{m}_0} \cdot d_{\tilde{m}_0} \tag{4.13}$$

$$q^{(0)}_{m_1} = \sum_{\tilde{m}_1=0}^{ncut} \left[T_{O_1}(\beta_1)\right]_{m_1 \tilde{m}_1} \cdot \hat{d}_{\tilde{m}_1} . \tag{4.14}$$

$[T_O(\beta_0)]_{m_0 \tilde{m}_0}$ and $\left[T_{O_1}(\beta_1)\right]_{m_1 \tilde{m}_1}$ are one of the T-matrices (2.5), (2.22), and (2.28), depending on the morphology of the corresponding sphere in its local system. It is important to pay attention to the dependence on the corresponding size parameter! $c^{(0)}_{m_0}$ and $q^{(0)}_{m_1}$ are nothing but the expansion coefficients of the scattered fields of the respective sphere in its local system in the absence of the other sphere. Introducing the more compact matrix-vector notation we may write instead of (4.11) and (4.12)

$$\vec{c}^{\,tp} = \vec{c}_0^{\,tp} - \mathbf{Q}_{12} \cdot \vec{q}^{\,tp} \tag{4.15}$$

and

$$\vec{q}^{\,tp} = \vec{q}_0^{\,tp} - \mathbf{C}_{21} \cdot \vec{c}^{\,tp} . \tag{4.16}$$

The elements of the matrices $\mathbf{Q}_{12}$ and $\mathbf{C}_{21}$ are given by

$$[Q_{12}]_{n n'} = -[T_O(\beta_0)]_{n n} \cdot (-1)^n \cdot S^0_{n n'}(k_0 b) , \; n, n' = 0, \ldots, ncut \tag{4.17}$$

and

$$[C_{21}]_{n n'} = -[T_{O_1}(\beta_1)]_{n n} \cdot (-1)^{n'} \cdot S^0_{n n'}(k_0 b) , \; n, n' = 0, \ldots, ncut . \tag{4.18}$$

$\vec{c}^{\,tp}$ and $\vec{q}^{\,tp}$ are the transposed vectors with the unknown expansion coefficients of the respective scattered field as their components. But before using this equation system to calculate the coefficients c_{m_0} and q_{m_1} we want to transform it further to end up with a consequent T-matrix picture of scattering on bispheres. This can be accomplished if using (4.16) to eliminate $\vec{q}^{\,tp}$ in (4.15), and if using (4.15) to eliminate $\vec{c}^{\,tp}$ in (4.16). This provides

$$\vec{c}^{\,tp} = \vec{c}_0^{\,tp} - \mathbf{Q}_{12} \cdot \vec{q}_0^{\,tp} + \mathbf{Q}_{12} \cdot \mathbf{C}_{21} \cdot \vec{c}^{\,tp} \tag{4.19}$$

and

$$\vec{q}^{\,tp} = \vec{q}_0^{\,tp} - \mathbf{C_{21}} \cdot \vec{c}_0^{\,tp} + \mathbf{C_{21}} \cdot \mathbf{Q_{12}} \cdot \vec{q}^{\,tp} \,. \tag{4.20}$$

An iteratively solvable T-matrix equation can be obtained from these two equations in the following way:

The "sphere-centered T-matrix" $\mathbf{T_s}$ of the bispheres, that links the known expansion coefficients of the incident plane wave to the expansion coefficients of the scattered fields in each local frame, is defined according to

$$\begin{pmatrix} \vec{c}^{\,tp} \\ \vec{q}^{\,tp} \end{pmatrix} := \mathbf{T_s} \cdot \begin{pmatrix} \vec{d}^{\,tp} \\ \vec{\hat{d}}^{\,tp} \end{pmatrix} \tag{4.21}$$

(see [8], for example). This allows us to transform (4.19) and (4.20) into the following single block matrix equation:

$$\mathbf{T_s} \cdot \vec{D}^{\,tp} = \mathbf{T_s^{(0)}} \cdot \vec{D}^{\,tp} - \mathbf{T_s^{(1)}} \cdot \vec{D}^{\,tp} +$$
$$\begin{pmatrix} \mathbf{Q_{12}} \cdot \mathbf{C_{21}} & \mathbf{O} \\ \mathbf{O} & \mathbf{C_{21}} \cdot \mathbf{Q_{12}} \end{pmatrix} \cdot \mathbf{T_s} \cdot \vec{D}^{\,tp} \,. \tag{4.22}$$

The block matrices therein are given by

$$\mathbf{T_s^{(0)}} = \begin{pmatrix} \mathbf{T_O} & \mathbf{O} \\ \mathbf{O} & \mathbf{T_{O_1}} \end{pmatrix} \tag{4.23}$$

and

$$\mathbf{T_s^{(1)}} = \begin{pmatrix} \mathbf{O} & \mathbf{Q_{12}} \cdot \mathbf{T_{O_1}} \\ \mathbf{C_{21}} \cdot \mathbf{T_O} & \mathbf{O} \end{pmatrix} \,, \tag{4.24}$$

and the block column vector reads

$$\vec{D}^{\,tp} = \begin{pmatrix} \vec{d}^{\,tp} \\ \vec{\hat{d}}^{\,tp} \end{pmatrix} \,. \tag{4.25}$$

$\mathbf{T_O}$ and $\mathbf{T_{O_1}}$ are again the diagonal T-matrices with elements given by $[T_O(\beta_0)]_{ii}$ and $[T_{O_1}(\beta_1)]_{ii}$. We thus end up with the T-matrix equation

$$\mathbf{T_s} = \mathbf{T_s^{(0)}} - \mathbf{T_s^{(1)}} + \begin{pmatrix} \mathbf{Q_{12}} \cdot \mathbf{C_{21}} & \mathbf{O} \\ \mathbf{O} & \mathbf{C_{21}} \cdot \mathbf{Q_{12}} \end{pmatrix} \cdot \mathbf{T_s} \tag{4.26}$$

that forms our starting point for an iterative solution. But before we can consider the iteration procedure in more detail we have to accomplish a final step:

Once we have determined the expansion coefficients of the scattered fields in each local system we must finally apply the inverse transformation of the scattered field of the shifted sphere into the laboratory frame to calculate the total scattering amplitude function. Eventually, we obtain

$$f_t(\theta) = -\frac{i}{k_0} \cdot \sum_{m_0=0}^{ncut} \left[\tilde{c}_{m_0} + \tilde{q}_{m_0} \right] \cdot (-i)^{m_0} \cdot Y_{0 m_0}(\theta, \phi = 0) , \qquad (4.27)$$

where

$$\begin{pmatrix} \vec{\tilde{c}}^{\,tp} \\ \vec{\tilde{q}}^{\,tp} \end{pmatrix} = \begin{pmatrix} \mathbf{E} & \mathbf{O} \\ \mathbf{O} & \mathbf{S_r} \end{pmatrix} \cdot \begin{pmatrix} \vec{c}^{\,tp} \\ \vec{q}^{\,tp} \end{pmatrix} \qquad (4.28)$$

are the expansion coefficients of the scattered fields in the laboratory frame. The elements of the matrix $\mathbf{S_r}$ are given by

$$[S_r]_{ij} = (-1)^i \cdot \widehat{S}_{ij}^0(k_0 b) ; \quad i, j = 0, \dots, ncut . \qquad (4.29)$$

$\mathbf{E}$ and $\mathbf{O}$ in the block matrix on the right-hand side of (4.28) represent the unit matrix and the zero matrix.

$$\begin{pmatrix} \vec{\tilde{c}}^{\,tp} \\ \vec{\tilde{q}}^{\,tp} \end{pmatrix} := \mathbf{T_{as}} \cdot \vec{D}^{\,tp} = \begin{pmatrix} \mathbf{E} & \mathbf{O} \\ \mathbf{O} & \mathbf{S_r} \end{pmatrix} \cdot \mathbf{T_s} \cdot \vec{D}^{\,tp} \qquad (4.30)$$

holds finally for the expansion coefficients of the scattered fields required in (4.27) to calculate the total scattering amplitude function in the laboratory frame. We will call $\mathbf{T_{as}}$ the T-matrix of axisymmetrically oriented bispheres in the laboratory frame.

Zero-Order Iteration

The zero-order iteration is obtained if only the first term on the right-hand side of (4.26) is taken into account. That is, we neglect any interaction between the spheres. Only the geometrical shift of the second sphere along the z-axis and the corresponding interference effect is considered. It is therefore of some interest to compare the corresponding results for the differential and total scattering cross-sections with the results of the approximation discussed in Sect. 4.1. First results for two identical sound soft spheres with increasing center distance b are presented in Figs. 4.7, 4.8 and 4.9. The following fixed parameters have been used in these computations: Radii: $a_0 = a_1 = 1.0$ mm, size parameters: $\beta_0 = \beta_1 = 3.0$, truncation parameter: $ncut = 9$. To our great pleasure we can state that the results of the approximation and the zero-order iteration of the more rigorous approach are exactly the same! This holds for the differential as well as the total scattering cross-sections with the latter being independent of the center distance b between the spheres. An assumption that is sometimes made if dilute ensembles of scatterers are considered is the assumption of independent scattering. It considers the scattering of each particle of the ensemble without reference to the other particles and neglects not only the interaction between the particles but also the interference due to the different positions of the particles. That a distance of three times the radius between the constituents of the ensemble is a sufficient condition to apply this approximation (see [9], Sect. 1.21 therein) is an often used estimate. Therefore, the corresponding results—simply the

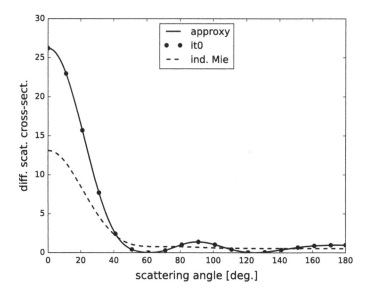

Fig. 4.7 Differential scattering cross-sections of an axisymmetrically oriented (s,s)-bisphere. Center distance: $b = 2.0$ mm (i.e., the spheres are touching!). Total scattering cross-sections: $\sigma_{tot} = 17.625$ (approxy), $\sigma_{tot} = 17.625$ (it0), $\sigma_{tot} = 8.813$ (Mie)

sum of the Mie results of the component spheres—are also plotted for intercomparison purposes. Regarding the above considered bispheres in a fixed axisymmetric orientation, it turns out that the differences are most pronounced in the forward direction while the independent scattering assumption seems to work quite well in the side- and backscattering region, where it looks like a least-squares fit of the approximation/zero-order iteration.

The same behavior can be observed if we replace the shifted sphere by a sound hard sphere but with the same radius and size parameter of the former sound soft sphere. The sound soft sphere that is centered in the laboratory frame remains unchanged. I.e., we have now the (s,h)-bisphere configuration. The corresponding results for the different center distances are presented in Figs. 4.10, 4.11 and 4.12. Increasing the size parameter of this latter configuration with the center distance of $b = 12$ mm to $\beta_0 = \beta_1 = 10$ shows that the region with the largest differences between the independent scattering assumption and the approximation is getting more and more concentrated in the forward direction (see Fig. 4.13). That the independent scattering assumption provides an acceptable accuracy at larger size parameters in the whole side- and backscattering region can be of some advantage for those practical applications that do not consider the forward scattering region.

Two More Sum Rules

At the end of Sect. 1.3.1 we have already discussed two sum rules resulting from the expansion of the incident plane wave and its transformation with respect to a shift along the z-axis of the laboratory frame, and for an arbitrary translation. The

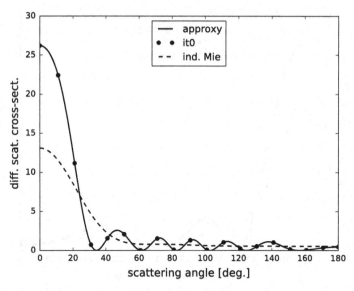

Fig. 4.8 Differential scattering cross-sections of an axisymmetrically oriented (s,s)-bisphere. Center distance: $b = 6.0$ mm. Total scattering cross-sections: $\sigma_{tot} = 17.625$ (approxy), $\sigma_{tot} = 17.625$ (it0)

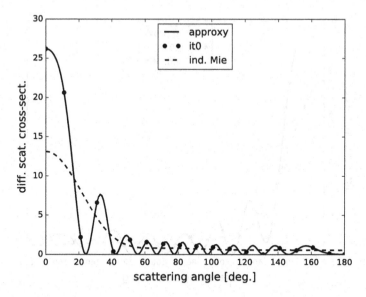

Fig. 4.9 Differential scattering cross-sections of an axisymmetrically oriented (s,s)-bisphere. Center distance: $b = 12.0$ mm. Total scattering cross-sections: $\sigma_{tot} = 17.625$ (approxy), $\sigma_{tot} = 17.625$ (it0)

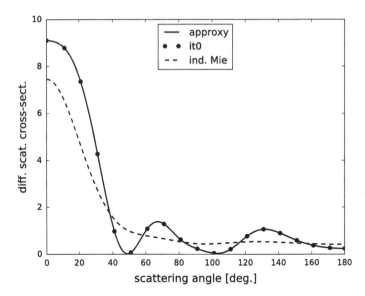

Fig. 4.10 Differential scattering cross-sections of an axisymmetrically oriented (s,h)-bisphere. Center distance: $b = 2.0$ mm (i.e., the spheres are touching!). Total scattering cross-sections: $\sigma_{tot} = 12.06$ (approxy), $\sigma_{tot} = 12.06$ (it0), $\sigma_{tot} = 8.813$ (Mie sound soft), $\sigma_{tot} = 3.247$ (Mie sound hard)

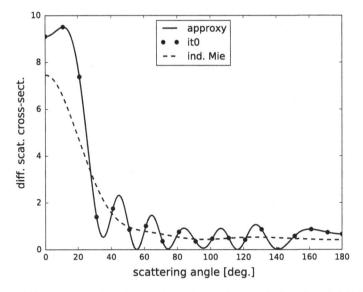

Fig. 4.11 Differential scattering cross-sections of an axisymmetrically oriented (s,h)-bisphere. Center distance: $b = 6.0$ mm. Total scattering cross-sections: $\sigma_{tot} = 12.06$ (approxy), $\sigma_{tot} = 12.06$ (it0)

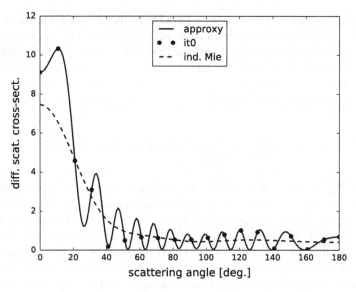

Fig. 4.12 Differential scattering cross-sections of an axisymmetrically oriented (s,h)-bisphere. Center distance: $b = 12.0$ mm. Total scattering cross-sections: $\sigma_{tot} = 12.06$ (approxy), $\sigma_{tot} = 12.06$ (it0)

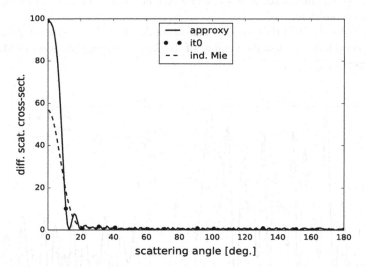

Fig. 4.13 Differential scattering cross-sections of an axisymmetrically oriented (s,h)-bisphere. Radii: $a_0 = a_1 = 1.0$ mm, size parameter: $\beta_0 = \beta_1 = 10.0$, truncation parameter: $ncut = 15$, center distance: $b = 12.0$ mm. Total scattering cross-sections: $\sigma_{tot} = 12.461$ (approxy), $\sigma_{tot} = 12.461$ (it0)

existence of another sum rule can be inferred from the observed numerical equivalence between the above considered zero-order iteration and the approximation of Sect. 4.1. Comparing the inverse transformation of the part of the scattering amplitude (4.27) that is related to the shifted sphere (note that this is accomplished by use of (4.28) and (4.29) to calculate $\vec{\tilde{q}}$!) with the corresponding part (4.1) of the simple approximation results in the relation

$$e^{-ik_0 b \cdot \cos\theta} \cdot Y_{0\,m_0}(\theta, \phi) = \sum_{m_1=0}^{ncut} i^{\,m_0+m_1} \cdot \widehat{S}^0_{m_0\,m_1}(k_0 b) \cdot Y_{0\,m_1}(\theta, \phi) \,. \tag{4.31}$$

The observed numerical equivalence is not a mathematical proof of this equivalence, of course! But it is straightforward to demonstrate at least the numerical correctness of the sum rule (4.31). A numerical test is shown in Figs. 4.14 and 4.15 for two different values of the truncation parameter $ncut$. An excellent agreement between the left- and the right-hand side of (4.31) can be observed if choosing $ncut = 40$. Please, note that only a few dots have been used to indicate the results of the left-hand side since no differences can be seen if using a full line. Applying this sum rule results in a simplification of the inverse transformation, and, therefore, in much less computational effort for axisymmetrically oriented bispheres. And, moreover, this simplification is independent of whether a certain order of iteration or the rigorous solution is considered.

Unfortunately, (4.31) holds only for the azimuthal mode $l = 0$. It seemed therefore quite natural to me to test the validity of the following generalization of (4.31) at least numerically:

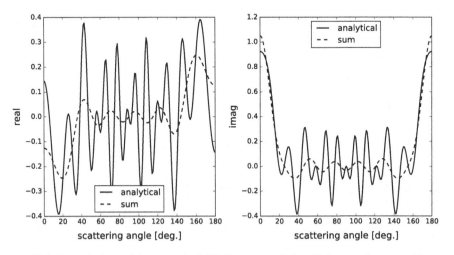

Fig. 4.14 Numerical test of the sum rule (4.31). Parameters: $k_0 b = 30.0$, $m_0 = 5$, $ncut = 15$

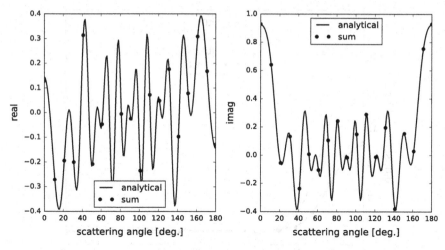

Fig. 4.15 Numerical test of the sum rule (4.31). Parameters: $k_0 b = 30.0$, $m_0 = 5$, $ncut = 40$

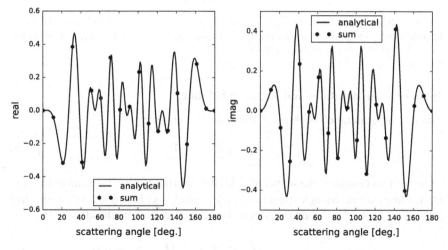

Fig. 4.16 Numerical test of the sum rule (4.32). Parameters: $kb = 30.0$, $m_0 = 5$, $l_1 = 2$, $ncut = 40$

$$e^{-ik_0 b \cdot \cos\theta} \cdot Y_{l_1 m_0}(\theta, \phi) = \sum_{m_1 = |l_1|}^{ncut} i^{m_0 + m_1} \cdot \widehat{S}^{l_1}_{m_0 m_1}(k_0 b) \cdot Y_{l_1 m_1}(\theta, \phi) . \qquad (4.32)$$

And the numerical test was indeed successful, as demonstrated with the result of Fig. 4.16. These two sum rules allow us again to test the correct numerical implementation of the separation matrix.

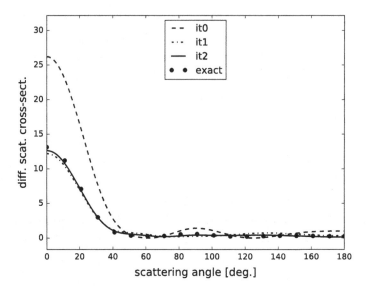

Fig. 4.17 Intercomparison of the different iterative and rigorous solutions for the bisphere configuration used in Fig. 4.7. Total scattering cross-sections: $\sigma_{tot} = 17.625$ (it0), $\sigma_{tot} = 11.638$ (it1), $\sigma_{tot} = 11.544$ (it2), $\sigma_{tot} = 11.410$ (rigorous)

Higher-Order Iterations and Rigorous Solution

The first-order iteration is obtained from (4.26) if choosing

$$\mathbf{T_s}^{(it1)} = \mathbf{T_s}^{(0)} - \mathbf{T_s}^{(1)} . \tag{4.33}$$

This iteration considers the scattered field, that is produced by the primary incident plane wave from each single sphere, as an additionally incident field for the respective other sphere. For the second-order iteration we get on the other hand

$$\mathbf{T_s}^{(it2)} = \mathbf{T_s}^{(it1)} + \begin{pmatrix} \mathbf{Q}_{12} \cdot \mathbf{C}_{21} & \mathbf{O} \\ \mathbf{O} & \mathbf{C}_{21} \cdot \mathbf{Q}_{12} \end{pmatrix} \cdot \mathbf{T_s}^{(it1)} . \tag{4.34}$$

It considers the scattered field of the first-order iteration of each sphere as the incident field for the respective other sphere, and so on. The rigorous solution follows from the inversion of

$$\begin{pmatrix} \mathbf{E} & \mathbf{Q}_{12} \\ \mathbf{C}_{21} & \mathbf{E} \end{pmatrix} \cdot \begin{pmatrix} \vec{c}^{\,tp} \\ \vec{q}^{\,tp} \end{pmatrix} = \begin{pmatrix} \mathbf{T_0} & \mathbf{O} \\ \mathbf{O} & \mathbf{T_{0_1}} \end{pmatrix} \cdot \begin{pmatrix} \vec{d}^{\,tp} \\ \vec{\lambda}^{\,tp} \end{pmatrix} . \tag{4.35}$$

This equation is identical with (4.15) and (4.16) but if both these equations are condensed into a block matrix notation!

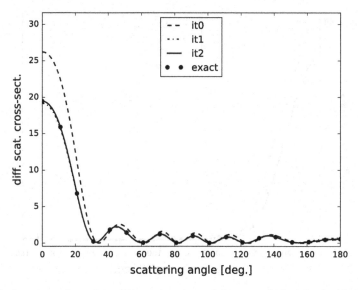

Fig. 4.18 Intercomparison of the different iterative and rigorous solutions for the bisphere configuration used in Fig. 4.8. Total scattering cross-sections: $\sigma_{tot} = 17.625$ (it0), $\sigma_{tot} = 13.770$ (it1), $\sigma_{tot} = 13.933$ (it2), $\sigma_{tot} = 13.935$ (rigorous)

To study the influence of the higher-order iterations and the rigorous solution in dependence on the center distance b let us go back to the bisphere configurations considered already in Figs. 4.7 and 4.8. The corresponding results are presented in Figs. 4.17 and 4.18. Looking at both Figs. we can state that the zero-order iteration produces the strongest differences especially in the forward region while it is not that bad in the side- and backscattering region, even for the touching spheres of Fig. 4.17. It is on the other hand remarkable that the differences in the differential as well as total scattering cross-sections between the different higher-order iterations and the rigorous solution of (4.35) are significantly lower as one may expect especially for the touching spheres. Regarding the higher-order iterations and the rigorous solution we have also to state that the total scattering cross-sections are now dependent on the center distance between the spheres. This is in contrast to what was observed for the zero-order iteration. We get a similar behavior if we replace the two sound soft spheres in Figs. 4.17 and 4.18 by two identical but sound penetrable spheres both with parameters $\kappa_k = 1.5$ and $\rho_p = 1.5$. All other parameters remain the same. The results are presented in Figs. 4.19 and 4.20. And what happens if we increase the size parameter further? The corresponding results for the configuration of Fig. 4.8 but at a larger size parameter of $\beta_0 = \beta_1 = 10.0$ are shown in Fig. 4.21. We can observe again the effect that the differences between the zero-order iteration and the other iterations and the rigorous solution are more and more concentrated in the forward direction while there are nearly no differences between the higher-order iterations and the rigorous solution.

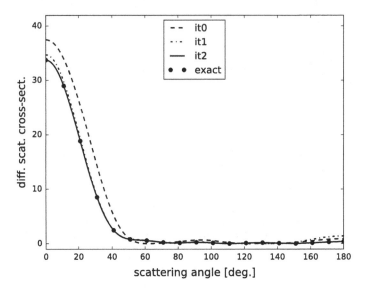

Fig. 4.19 Intercomparison of the different iterative and rigorous solutions for a (p,p)-bisphere configuration at a size parameter and a center distance that was used in Fig. 4.7. Total scattering cross-sections: $\sigma_{tot} = 22.050$ (it0), $\sigma_{tot} = 23.626$ (it1), $\sigma_{tot} = 23.360$ (it2), $\sigma_{tot} = 23.363$ (rigorous)

All the scattering computations of this section have been performed with the Python program *bisphere_z_oriented.py* that can be found in Appendix D. We will now abandon the restriction to axisymmetrically oriented bispheres.

4.2.2 Arbitrary Orientation

To derive the T-matrix equation (or, better, the T-operator equation) for an arbitrarily oriented bisphere configuration we can choose the same way used in the previous section, in principle. But the sphere that was so far only shifted along the z-axis of the laboratory frame is now located somewhere in this system while the other sphere remains centered. Accordingly, the transformation behavior of the scattered fields produced by each sphere in its local system becomes a little bit more complicate. In Sect. 1.2.3 we have already dealt with two different ways of such a transformation. Since restricting the considerations to homogeneous spheres with homogeneous boundary conditions on its whole surface (i.e., since we do not consider combinations of nonspherical objects or Janus spheres) we can benefit from the second transformation—the translation of the laboratory frame according to Fig. 1.12. The advantages of using this transformation are the following:

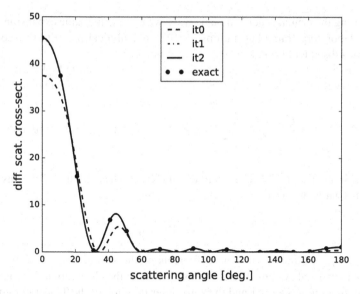

Fig. 4.20 Intercomparison of the different iterative and rigorous solutions for a (p,p)-bisphere configuration at a size parameter and a center distance that was used in Fig. 4.8. Total scattering cross-sections: $\sigma_{tot} = 22.050$ (it0), $\sigma_{tot} = 27.052$ (it1), $\sigma_{tot} = 27.118$ (it2), $\sigma_{tot} = 27.118$ (rigorous)

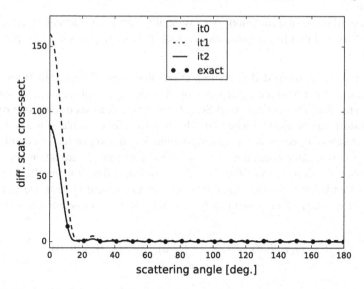

Fig. 4.21 Intercomparison of the different iterative and rigorous solutions for the bisphere configuration used in Fig. 4.8 but for a size parameter of $\beta_0 = \beta_1 = 10.0$ and a truncation parameter of $ncut = 15$. Total scattering cross-sections: $\sigma_{tot} = 15.062$ (it0), $\sigma_{tot} = 10.173$ (it1), $\sigma_{tot} = 10.255$ (it2), $\sigma_{tot} = 10.255$ (rigorous)

- Regarding the primary incident plane wave we can use the same expansion known in the laboratory frame but multiplied with an analytical phase term (see (1.81) and the subsequent remark).
- The transformation of the scattered field generated by the shifted sphere back to the laboratory frame to calculate the coefficients $\tilde{q}_{l_0 m_0}$ in the scattering phase function

$$f_t(\theta, \phi) = -\frac{i}{k_0} \cdot \sum_{l_0=-m_0}^{m_0} \sum_{m_0=0}^{ncut} \left[\tilde{c}_{l_0 m_0} + \tilde{q}_{l_0 m_0} \right] \cdot (-i)^{m_0} \cdot Y_{l_0 m_0}(\theta, \phi) \quad (4.36)$$

can again be accomplished if using the analytical phase term $e^{-ik_0\vec{b}\cdot\hat{r}_\theta}$ instead of (1.62). That is, we have

$$\tilde{q}_{l_0 m_0} = e^{-ik_0\vec{b}\cdot\hat{r}_\theta} \cdot q_{l_0 m_0}, \quad (4.37)$$

where $q_{l_0 m_0}$ are the expansion coefficients of the scattered field of the shifted sphere in its local system. This is a consequence of the observation that the simple approximation of Sect. 4.1 and the zero-order iteration of the T-operator equation we will derive in this section produce again identical scattering cross-sections. The zero-order iteration requires the more complicated back transformation according to (1.62).

- Because of these simplifications, we eventually obtain a numerical procedure that can be applied to a larger parameter region, and that requires less computational effort.

However, if one is interested in configurations that consist of combinations of two nonspherical but rotational symmetric objects or Janus spheres as discussed in Chap. 3, the other transformation of Sect. 1.2.3, which consists of only one rotation and one shift, can be used with benefit. This happens, for example, if the new z-axis after the rotation agrees with the axis of rotational symmetry of such a combination.

Only the boundary conditions for two sound soft spheres are again considered in the following derivation of the T-operator equation. But, if using the respective T-matrix that belongs to each component sphere in its local system, we can apply this equation to any other combination of sound soft, hard, or penetrable spheres.

$$u_{s_0}(k_0 r, \theta, \phi) = \sum_{l_0=-m_0}^{m_0} \sum_{m_0=0}^{ncut} c_{l_0 m_0} \cdot \varphi_{l_0 m_0}(k_0 r, \theta, \phi) \quad (4.38)$$

and

$$u_{s_1}(k_0 r_1, \theta_1, \phi_1) = \sum_{l_1=-m_1}^{m_1} \sum_{m_1=0}^{ncut} q_{l_1 m_1} \cdot \varphi_{l_1 m_1}(k_0 r_1, \theta_1, \phi_1) \quad (4.39)$$

are now the expansions of the scattered fields in the local system of each sphere. Applying the boundary condition (2.1) for a sound soft sphere in each of the local systems, and if taking the transformation relations (1.58) and (1.61) of a translation of the laboratory frame into account result in the following two equations:

- In the laboratory frame $\mathbf{O}$ at $r = a_0$:

$$\sum_{l_0=-m_0}^{m_0} \sum_{m_0=0}^{ncut} c_{l_0 m_0} \cdot \varphi_{l_0 m_0}(\beta_0, \theta, \phi) = -\sum_{m_0=0}^{ncut} d_{m_0} \cdot \psi_{0 m_0}(\beta_0, \theta) -$$

$$\sum_{l_1=-m_1}^{m_1} \sum_{m_1=0}^{ncut} \sum_{l_1'=-m_1'}^{m_1'} \sum_{m_1'=0}^{ncut} q_{l_1 m_1} \cdot \left[T_{rsr}^{-1}\right]_{m_1 m_1'}^{l_1 l_1'} \cdot \psi_{l_1' m_1'}(\beta_0, \theta, \phi) \, .$$

(4.40)

- In the local system $\mathbf{O}_1$ of the shifted sphere at $r = a_1$:

$$\sum_{l_1=-m_1}^{m_1} \sum_{m_1=0}^{ncut} q_{l_1 m_1} \cdot \varphi_{l_1 m_1}(\beta_1, \theta_1, \phi_1) = -e^{-ik_0 b \cos \theta_p} \, .$$

$$\sum_{m_1=0}^{ncut} d_{m_1} \cdot \psi_{0 m_1}(\beta_1, \theta_1) - \sum_{l_0=-m_0}^{m_0} \sum_{m_0=0}^{ncut} \sum_{l_0'=-m_0'}^{m_0'} \sum_{m_0'=0}^{ncut} c_{l_0 m_0} \cdot$$

$$\left[T_{rsr}\right]_{m_0 m_0'}^{l_0 l_0'} \cdot \psi_{l_0' m_0'}(\beta_1, \theta_1, \phi_1) \, .$$

(4.41)

d_{m_0} and d_{m_1} are the expansion coefficients of the incident plane wave according to (1.70). These equations are next multiplied by the spherical harmonics $Y_{\tilde{l}_0 \tilde{m}_0}(\theta, \phi)$ or $Y_{\tilde{l}_0 \tilde{m}_0}(\theta_1, \phi_1)$ and integrated over the respective spherical surface. In so doing, and due to the orthonormality relation (1.14) we now end up with the equation system

$$c_{l_0 m_0} - [T_O(\beta_0)]_{m_0 m_0} \cdot \sum_{l_1=-m_1}^{m_1} \sum_{m_1=0}^{ncut} \left[T_{rsr}^{-1}\right]_{m_1 m_0}^{l_1 l_0} \cdot q_{l_1 m_1} = c_{l_0 m_0}^{(0)}$$

(4.42)

and

$$q_{l_1 m_1} - [T_{O_1}(\beta_1)]_{m_1 m_1} \cdot \sum_{l_0=-m_0}^{m_0} \sum_{m_0=0}^{ncut} \left[T_{rsr}\right]_{m_0 m_1}^{l_0 l_1} \cdot c_{l_0 m_0} = q_{l_1 m_1}^{(0)}$$

(4.43)

to determine the expansion coefficients $c_{l_0 m_0}$ and $q_{l_1 m_1}$ of the scattered fields in each local system. This is the system analogous to (4.11) and (4.12), and $c_{l_0 m_0}^{(0)}$ and $q_{l_1 m_1}^{(0)}$ therein are given by

$$c_{l_0 m_0}^{(0)} = \sum_{l'=-n'}^{n'} \sum_{n'=0}^{ncut} [T_O(\beta_0)]_{m_0 n'}^{l_0 l'} \cdot d_{l' n'}$$

(4.44)

$$q_{l_1 m_1}^{(0)} = \sum_{l'=-n'}^{n'} \sum_{n'=0}^{ncut} \left[T_{O_1}(\beta_1) \right]_{m_1 n'}^{l_1 l'} \cdot \widehat{d}_{l' n'} . \tag{4.45}$$

$d_{l' n'}$ and $\widehat{d}_{l' n'}$ are the known expansion coefficients of the incident field according to (2.6), and

$$\widehat{d}_{l' n'} = e^{ik_0 b \cdot \cos \theta_p} \cdot d_{l' n'} . \tag{4.46}$$

Depending on the morphology of the respective sphere in its local system, the T-matrices $\left[T_O(\beta_0) \right]_{m_0 n'}^{l_0 l'}$ and $\left[T_{O_1}(\beta_1) \right]_{m_1 n'}^{l_1 l'}$ are identical with (2.5), (2.22), or (2.28). I.e., (4.44) and (4.45) are again the solutions for each single sphere in its local system if neglecting the interaction between the spheres.

To proceed as in the previous section let us next introduce the two quantities $\mathcal{Q}_{12}$ and $\mathcal{C}_{21}$ with elements according to

$$[Q_{12}]_{n n'}^{l l'} = - [T_O(\beta_0)]_{n n} \cdot \left[T_{rsr}^{-1} \right]_{n' n}^{l' l} \tag{4.47}$$

$$[C_{21}]_{n n'}^{l l'} = - e^{ik_0 b \cdot \cos \theta_p} \cdot \left[T_{O_1}(\beta_1) \right]_{n n} \cdot [T_{rsr}]_{n' n}^{l' l} . \tag{4.48}$$

We will call these quantities "operators" in the subsequent discussion to distinguish them from the conventional matrices which appear in the previous section. Let us furthermore define the "product" of two such operators by

$$\mathcal{A} = \mathcal{B} \cdot \mathcal{C} := \sum_{\bar{l}=-\bar{n}}^{\bar{n}} \sum_{\bar{n}=0}^{ncut} [\mathcal{B}]_{n \bar{n}}^{l \bar{l}} \cdot [\mathcal{C}]_{\bar{n} n'}^{\bar{l} l'} = [\mathcal{A}]_{n n'}^{l l'} , \tag{4.49}$$

and the product $\mathbf{B} = \mathcal{A} \cdot \mathbf{A}$ of an operator $\mathcal{A}$ with a coefficient matrix $\mathbf{A}$ (i.e., the expansion coefficients are now expressed by a conventional matrix $\mathbf{A}$ with elements $[A]_{l n}$) by

$$[B]_{l n} := \sum_{l'=-n'}^{n'} \sum_{n'=0}^{ncut} [\mathcal{A}]_{n n'}^{l l'} \cdot [A]_{l' n'} . \tag{4.50}$$

This allows us to condense (4.42)–(4.45) into the block operator equation

$$\begin{pmatrix} \mathcal{E} & \mathcal{Q}_{12} \\ \mathcal{C}_{21} & \mathcal{E} \end{pmatrix} \cdot \begin{pmatrix} \mathbf{C} \\ \mathbf{Q} \end{pmatrix} = \begin{pmatrix} \mathbf{C_0} \\ \mathbf{Q_0} \end{pmatrix} = \begin{pmatrix} \mathcal{T}_O & \mathcal{O} \\ \mathcal{O} & \mathcal{T}_{O_1} \end{pmatrix} \cdot \begin{pmatrix} \mathbf{D} \\ \widehat{\mathbf{D}} \end{pmatrix} . \tag{4.51}$$

This is the expression analogous to (4.35). $\mathcal{E}$ and $\mathcal{O}$ represent the unit operator and the zero operator.

Looking at (4.21), the "sphere-centered T-operator" $\mathcal{T}_s$ of the bispheres can now be defined according to

$$\begin{pmatrix} \mathbf{C} \\ \mathbf{Q} \end{pmatrix} := \mathcal{T}_s \cdot \begin{pmatrix} \mathbf{D} \\ \hat{\mathbf{D}} \end{pmatrix} . \tag{4.52}$$

Following again the procedure applied in the previous section we thus end up with

$$\mathcal{T}_s = \mathcal{T}_s^{(0)} - \mathcal{T}_s^{(1)} + \begin{pmatrix} \mathcal{Q}_{12} \cdot \mathcal{C}_{21} & \mathcal{O} \\ \mathcal{O} & \mathcal{C}_{21} \cdot \mathcal{Q}_{12} \end{pmatrix} \cdot \mathcal{T}_s , \tag{4.53}$$

where

$$\mathcal{T}_s^{(0)} = \begin{pmatrix} \mathcal{T}_\mathbf{O} & \mathcal{O} \\ \mathcal{O} & \mathcal{T}_{\mathbf{O}_1} \end{pmatrix} \tag{4.54}$$

and

$$\mathcal{T}_s^{(1)} = \begin{pmatrix} \mathcal{O} & \mathcal{Q}_{12} \cdot \mathcal{T}_{\mathbf{O}_1} \\ \mathcal{C}_{21} \cdot \mathcal{T}_\mathbf{O} & \mathcal{O} \end{pmatrix} . \tag{4.55}$$

Equation (4.53) forms again the starting point for an iterative solution of the scattering problem of arbitrarily oriented bisphere configurations. The corresponding numerical procedures up to the second-order iteration is implemented in the Python program *bisphere_it_trans*.

Zero-Order Iteration

The zero-order iteration is again obtained if only the first term on the right-hand side of (4.53) is taken into account. To compare the differential and total scattering cross-sections with the results of the approximation discussed in Sect. 4.1 we consider in a first example the same configuration of two identical sound soft spheres used in Figs. 4.7, 4.8 and 4.9 (just to remember: radii: $a_0 = a_1 = 1.0$ mm, size parameter: $\beta_0 = \beta_1 = 3.0$, truncation parameter: $ncut = 9$). But the second sphere is now located on the x-axis of the laboratory frame. I.e., the Eulerian angles ($\alpha = 0°, \theta_p = 90°$) are now used for all the three different center distances. Since this configuration consists of two identical sound soft spheres it is sufficient to plot the results only in the interval $\theta \in [0, \pi]$. This is due to the symmetry of the differential scattering cross-sections with respect to the z-axis. Looking at Figs. 4.22, 4.23 and 4.24 we can again find an excellent match between the approximation and the zero-order iteration. Regarding the differences in the differential scattering cross-sections between the two different orientations used in Figs. 4.7, 4.8 and 4.9 and Figs. 4.22, 4.23 and 4.24 we can observe major differences especially in the forward direction. The forward peaks become much narrower in the latter cases. But there are also differences in the ripple structure in the side and back scattering region. Now we are also able to test the reciprocity condition. To this end, let us consider the two orientations depicted in Fig. 1.13 (see Sect. 1.4). That is, we compare the differential scattering cross-sections of the axisymmetrically oriented bispheres of Figs. 4.7, 4.8 and 4.9 (there we have $\theta_p = 0°$) at a scattering angle of $\theta = 90°$ with the differential scattering cross-sections of the bispheres of Figs. 4.22, 4.23 and 4.24 (here we have

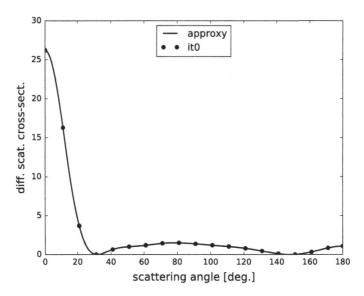

Fig. 4.22 Differential scattering cross-sections of a (s,s)-bisphere configuration with the second sphere located on the x-axis of the laboratory frame. Center distance: $b = 2.0$ mm (touching spheres!). Total scattering cross-sections: $\sigma_{\text{tot}} = 17.625$ (approxy), $\sigma_{\text{tot}} = 17.625$ (it0). Differential scattering cross-sections: $\partial\sigma/\partial\Omega = 1.4068$ ($\theta_p = 0°$), $\partial\sigma/\partial\Omega = 1.4068$ ($\theta_p = 90°$)

$\theta_p = 90°$) at the same scattering angle of $\theta = 90°$ (this is possible because of the symmetry of these configurations!). The corresponding values are given in the figure captions of Figs. 4.22, 4.23 and 4.24. The reciprocity condition is obviously well fulfilled. That this condition is already met on the level of the zero-order iteration raises the question if it can be used to estimate the overall accuracy of an obtained scattering solution for bispheres. It seems that this is not possible since the higher order iterations end up with different values for the respective differential scattering cross-sections. This will be demonstrated in the next subsection. However, it turns out that this condition is always fulfilled if we compare the differential scattering cross-sections that result from the same level of iteration. This provides us with a further possibility to test the correct numerical implementation of the iteration procedure.

The second example is concerned with the bisphere configuration of Figs. 4.10, 4.11 and 4.12 but with the sound hard sphere now shifted along the x-axis of the laboratory frame. The match between the approximation and the zero-order iteration is again excellent, as one can see from Figs. 4.25, 4.26 and 4.27. The same holds for the fulfillment of the reciprocity condition. But due to this asymmetric bisphere configuration and the resulting asymmetric behavior of the differential scattering cross-section we must now compare the differential scattering cross-sections of the axisymmetrically oriented bispheres (i.e., $\theta_p = 0°$) at the scattering angle of $\theta = 90°$ with the differential scattering cross-sections of the corresponding configurations with the sound hard sphere shifted along the x-axis (i.e., $\theta_p = 90°$) at the scattering

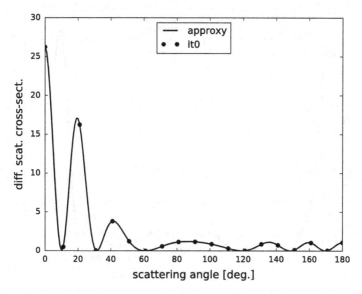

Fig. 4.23 Differential scattering cross-sections of a (s,s)-bisphere configuration with the second sphere located on the x-axis of the laboratory frame. Center distance: $b = 6.0$ mm. Total scattering cross-sections: $\sigma_{tot} = 17.625$ (approxy), $\sigma_{tot} = 17.625$ (it0). Differential scattering cross-sections: $\partial\sigma/\partial\Omega = 1.1916$ ($\theta_p = 0°$), $\partial\sigma/\partial\Omega = 1.1916$ ($\theta_p = 90°$)

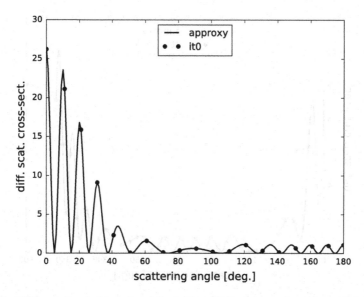

Fig. 4.24 Differential scattering cross-sections of a (s,s)-bisphere configuration with the second sphere located on the x-axis of the laboratory frame. Center distance: $b = 12.0$ mm. Total scattering cross-sections: $\sigma_{tot} = 17.625$ (approxy), $\sigma_{tot} = 17.625$ (it0). Differential scattering cross-sections: $\partial\sigma/\partial\Omega = 0.6259$ ($\theta_p = 0°$), $\partial\sigma/\partial\Omega = 0.6259$ ($\theta_p = 90°$)

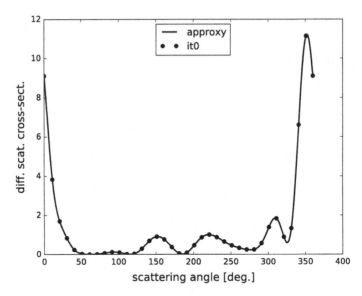

Fig. 4.25 Differential scattering cross-sections of a (s,h)-bisphere configuration with the sound hard sphere located on the x-axis of the laboratory frame. Center distance: $b = 2.0$ mm (touching spheres!). Total scattering cross-sections: $\sigma_{\text{tot}} = 12.06$ (approxy), $\sigma_{\text{tot}} = 12.06$ (it0). Differential scattering cross-sections: $\partial\sigma/\partial\Omega = 0.2597$ ($\theta_p = 0°$), $\partial\sigma/\partial\Omega = 0.2597$ ($\theta_p = 90°$)

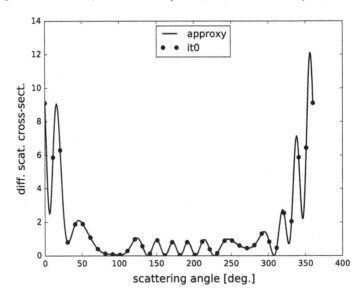

Fig. 4.26 Differential scattering cross-sections of a (s,h)-bisphere configuration with the sound hard sphere located on the x-axis of the laboratory frame. Center distance: $b = 6.0$ mm. Total scattering cross-sections: $\sigma_{\text{tot}} = 12.06$ (approxy), $\sigma_{\text{tot}} = 12.06$ (it0). Differential scattering cross-sections: $\partial\sigma/\partial\Omega = 0.4595$ ($\theta_p = 0°$), $\partial\sigma/\partial\Omega = 0.4595$ ($\theta_p = 90°$)

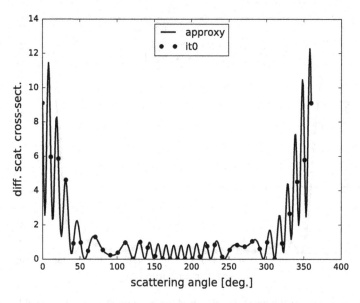

Fig. 4.27 Differential scattering cross-sections of a (s,h)-bisphere configuration with the sound hard sphere located on the x-axis of the laboratory frame. Center distance: $b = 12.0$ mm. Total scattering cross-sections: $\sigma_{tot} = 12.06$ (approxy), $\sigma_{tot} = 12.06$ (it0). Differential scattering cross-sections: $\partial\sigma/\partial\Omega = 0.7366$ ($\theta_p = 0°$), $\partial\sigma/\partial\Omega = 0.7366$ ($\theta_p = 90°$)

angle of $\theta = 270°$. These values are again given in the figure captions. And also the slight halo effect to be seen in Figs. 4.25, 4.26 and 4.27 should not go unmentioned. This effect is shifted more and more toward the forward direction for an increasing center distance.

Higher-Order Iterations

In this subsection we want to examine the effect of the higher order iterations on the total and differential scattering cross-sections. For this purpose we repeat the computations for the above two configurations, but now for the zero-, first- and second-order iteration. The results for the configuration of the two sound soft spheres are presented in Figs. 4.28, 4.29 and 4.30. Comparing these results with the results of Figs. 4.17 and 4.18 of the axisymmetrically oriented bispheres one can clearly see that the correspondence between the different orders of iteration has improved. Especially the differences between the zero- and first-order iteration have become much smaller and even more focused on the forward direction. Differences between the first- and second-order iteration are only hard to see. This behavior can be explained as follows: Regarding the axisymmetric orientation, the one sphere is shifted in the forward direction. But this is the direction with the strongest intensity of the scattered field produced by the centered sphere. As a consequence, one may expect a greater influence of the interaction contribution on the cross-sections for this orientation. On the other hand, if the shifted sphere is located on the x-axis of the laboratory frame,

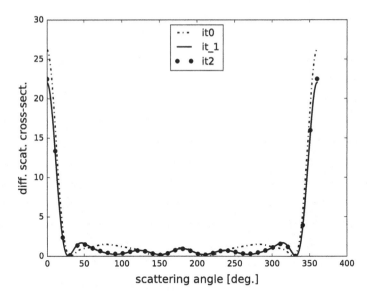

Fig. 4.28 Differential scattering cross-sections of the (s,s)-bisphere configuration of Fig. 4.22 but for different orders of iteration. Total scattering cross-sections: $\sigma_{\text{tot}} = 17.625$ (it0), $\sigma_{\text{tot}} = 16.302$ (it1), $\sigma_{\text{tot}} = 16.488$ (it2)

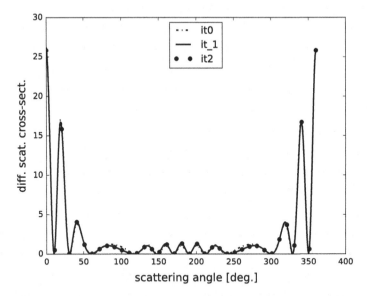

Fig. 4.29 Differential scattering cross-sections of the (s,s)-bisphere configuration of Fig. 4.23 but for different orders of iteration. Total scattering cross-sections: $\sigma_{\text{tot}} = 17.625$ (it0), $\sigma_{\text{tot}} = 17.211$ (it1), $\sigma_{\text{tot}} = 17.20$ (it2)

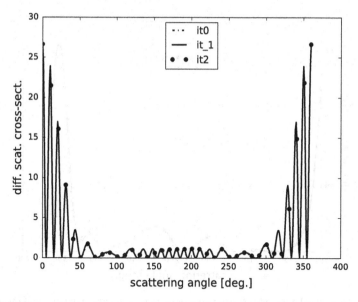

Fig. 4.30 Differential scattering cross-sections of the (s,s)-bisphere configuration of Fig. 4.24 but for different orders of iteration. Total scattering cross-sections: $\sigma_{tot} = 17.625$ (it0), $\sigma_{tot} = 17.66$ (it1), $\sigma_{tot} = 17.656$ (it2)

the scattered field of the centered sphere has usually a much lower intensity in this direction. This should result in a reduction of the interaction effect, as confirmed by the results. Looking at Figs. 4.31, 4.32 and 4.33 we can observe the same behavior for the (s,h)-bisphere configuration. Figure 4.31 reveals moreover that the halo effect has survived.

From the calculations made so far in this chapter results an interesting conclusion: It seems that already the first-order iteration produces results of a sufficient accuracy for many practical applications.

Ensembles of Bispheres

In this last subsection we abandon the restriction to bispheres in fixed orientations. For this purpose we will present a few results for ensembles of identical but independently scattering bispheres. The considered ensembles consists of differently oriented bispheres while the size parameter and the center distance remain fixed. The different orders of iteration are compared against the cross-sections that result from the independent scattering assumption discussed in Sect. 4.2.1. Due to the symmetry of the considered configurations we can re-strict the computations for both Eulerian angles to the interval [0°, 90°]. This interval is covered in steps of 10°. That is, the results for each order of iteration in Figs. 4.34, 4.35, 4.36, 4.37, 4.38 and 4.39 represents an average over 100 different orientations. This procedure is of course not a random orientation of bispheres but prefers the forward direction.

The most obvious result of these computations is the fact that the differences between the differential scattering cross-sections of the independent scattering

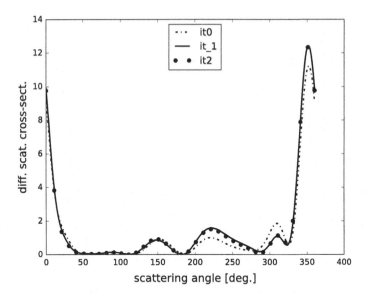

Fig. 4.31 Differential scattering cross-sections of the (s,h)-bisphere configuration of Fig. 4.25 but for different orders of iteration. Total scattering cross-sections: $\sigma_{tot} = 12.06$ (it0), $\sigma_{tot} = 12.73$ (it1), $\sigma_{tot} = 12.704$ (it2)

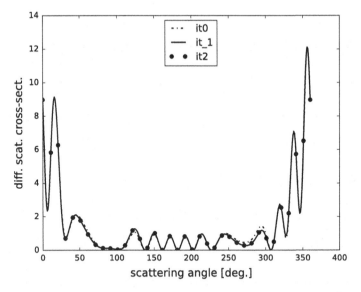

Fig. 4.32 Differential scattering cross-sections of the (s,h)-bisphere configuration of Fig. 4.26 but for different orders of iteration. Total scattering cross-sections: $\sigma_{tot} = 12.06$ (it0), $\sigma_{tot} = 12.062$ (it1), $\sigma_{tot} = 12.043$ (it2)

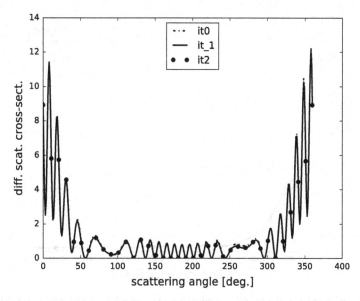

Fig. 4.33 Differential scattering cross-sections of the (s,h)-bisphere configuration of Fig. 4.27 but for different orders of iteration. Total scattering cross-sections: $\sigma_{tot} = 12.06$ (it0), $\sigma_{tot} = 11.951$ (it1), $\sigma_{tot} = 11.954$ (it2)

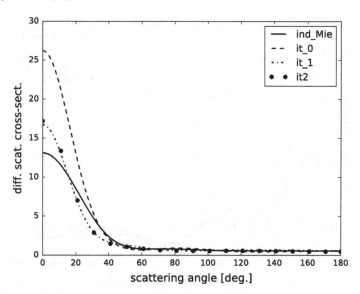

Fig. 4.34 Differential scattering cross-sections of a (s,s)-bisphere configuration in random orientation with respect to the Eulerian angles (α, θ_p). Parameters: radii $a_0 = a_1 = 1.0$ mm, size parameters $\beta_0 = \beta_1 = 3.0$, center distance $b = 2.0$ mm (touching spheres), truncation parameter $ncut = 11$

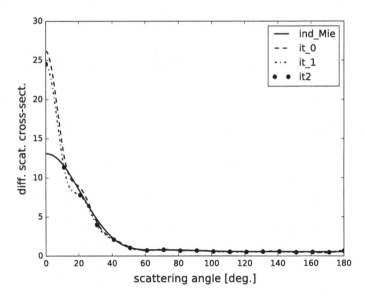

Fig. 4.35 Differential scattering cross-sections of a (s,s)-bisphere configuration in random orientation with respect to the Eulerian angles (α, θ_p). Parameters: radii $a_0 = a_1 = 1.0$ mm, size parameters $\beta_0 = \beta_1 = 3.0$, center distance $b = 6.0$ mm (touching spheres), truncation parameter $ncut = 11$

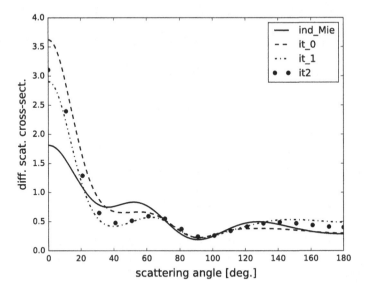

Fig. 4.36 Differential scattering cross-sections of a (h,h)-bisphere configuration in random orientation with respect to the Eulerian angles (α, θ_p). Parameters: radii $a_0 = a_1 = 1.0$ mm, size parameters $\beta_0 = \beta_1 = 3.0$, center distance $b = 2.0$ mm (touching spheres), truncation parameter $ncut = 11$

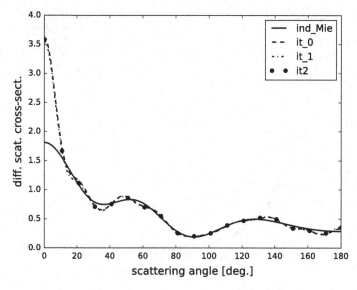

Fig. 4.37 Differential scattering cross-sections of a (h,h)-bisphere configuration in random orientation with respect to the Eulerian angles (α, θ_p). Parameters: radii $a_0 = a_1 = 1.0$ mm, size parameters $\beta_0 = \beta_1 = 3.0$, center distance $b = 6.0$ mm (touching spheres), truncation parameter $ncut = 11$

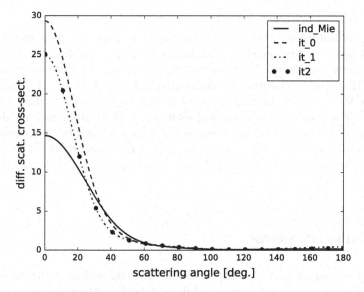

Fig. 4.38 Differential scattering cross-sections of a (p,p)-bisphere configuration in random orientation with respect to the Eulerian angles (α, θ_p). Parameters: radii $a_0 = a_1 = 1.0$ mm, size parameters $\beta_0 = \beta_1 = 3.0$, $\kappa_k = 1.5$, $\rho_p = 2.25$, center distance $b = 2.0$ mm (touching spheres), truncation parameter $ncut = 11$

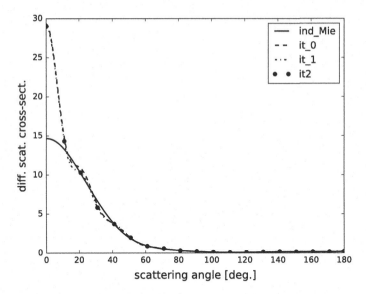

Fig. 4.39 Differential scattering cross-sections of a (p,p)-bisphere configuration in random orientation with respect to the Eulerian angles (α, θ_p). Parameters: radii $a_0 = a_1 = 1.0$ mm, size parameters $\beta_0 = \beta_1 = 3.0$, $\kappa_k = 1.5$, $\rho_p = 2.25$, center distance $b = 6.0$ mm (touching spheres), truncation parameter $ncut = 11$

assumption and the different orders of iteration are again most pronounced in the forward region, and independent of the center distance b. But these differences are more and more concentrated to the forward direction $\theta = 0°$ for increasing center distances. This behavior confirms the statement made by van de Hulst in [9] (Sect. 1.21 therein). Moreover, we can state again that already the first-order iteration produces quite accurate results for many practical applications. This is independent of the center distance. The corresponding numerical procedure can easily be implemented and is quite fast. However, it seems that for larger center distances results of a sufficient accuracy are already obtained with the zero-order iteration/the simple approximation!

4.3 Python Programs

This section contains three types of programs based on the methods we have discussed in this chapter to calculate the scattering behavior of different bisphere configurations. These are the approximation that neglects the interaction between the spheres, the iterative and rigorous solution of the T-matrix equation for axisymmetrically oriented bispheres, and the iterative solution of the T-operator method for arbitrarily oriented bispheres. They are organized in the same way as described at the beginning of

Sect. 2.5. That is, the respective program that generates the input file must be started first. A program to test the two sum rules (4.31) and (4.32) is also included.

The Python programs of this chapter are the following:

- programs *bisphere_approxy.py* and *bisphere_approxy_input.py*:
  These programs can be used to calculate the differential and total scattering cross-sections of any bisphere combination of sound soft, hard, or penetrable spheres. The bispheres can be arbitrarily oriented in the laboratory frame. Only the phase difference that results from the different geometrical positions of the spheres in the laboratory frame is considered. Any interaction between the spheres is neglected. The computation is performed at a given size parameter that is related to the centered sphere.
- programs *bisphere_z_oriented.py* and *bisphere_z_oriented_input.py*:
  These programs compute the differential and total scattering cross-sections of any bisphere combination of sound soft, hard, or penetrable spheres. But only bispheres that are axisymmetrically oriented in the laboratory frame can be analyzed. The scattering problem is solved by an iteration procedure that is applied to the T-matrix equation (4.26), or by a rigorous solution of (4.35). The computation is performed at a given size parameter that is related to the centered sphere, and with a truncation parameter *ncut* that must be specified by the user.
- programs *bisphere_it_trans.py* and *bisphere_it_trans_input.py*: These programs compute the differential and total scattering cross-sections of any bisphere combination of sound soft, hard, or penetrable spheres. The bispheres can be arbitrarily oriented in the laboratory frame. The scattering problem is solved by an iteration procedure that is applied to the T-operator equation (4.53). The computation is performed at a given size parameter that is related to the centered sphere, and with a truncation parameter *ncut* that must be specified by the user.
- program *sumtest_1.py*: This program can be used to test the two sum rules (4.31) and (4.32).

The full programs are given in Appendix D.

References

1. Eyges, L.: Some nonseparable boundary value problems and the many-body problem. Ann. Phys. **2**, 101–128 (1957)
2. Peterson, B.: Numerical computations of acoustic scattering from two spheres. Swedish Institute of Applied Mathematics, Sweden, Internal Report T.M.F. 73-2 (1973)
3. Kapodistrias, G., Dahl, P.H.: Effects of interaction between two bubble scatterers. JASA **107**, 3006–3017 (2000)
4. Gaunaurd, G.C., Huang, H., Strifors, H.C.: Acoustic scattering by a pair of spheres. JASA **98**, 495–507 (1995)
5. Martin, P.A.: Multiple Scattering: Interaction of Time-Harmonic Waves with N Obstacles. Cambridge University Press, Cambridge (UK) (2006)
6. Mishchenko, M.I., Mackowski, D.W., Travis, L.D.: Scattering of light by bispheres with touching and separated components. Appl. Opt. **34**, 4589–4599 (1995)

7. Rother, T.: Green's Functions in Classical Physics. Springer International Publishing AG, Cham, Switzerland (2017)
8. Mishchenko, M.I., Mackowski, D.W.: Light scattering by randomly oriented bispheres. Opt. Lett. **19**, 1604–1606 (1994)
9. van de Hulst, H.C.: Light Scattering by Small Particles. Dover, New York (1981)

Appendix A
Full Python Programs of Chap. 1

© Springer Nature Switzerland AG 2020

T. Rother, *Sound Scattering on Spherical Objects*,

https://doi.org/10.1007/978-3-030-36448-9

```
#        Modul "basics.py"
#
# This module containes:
#
# subroutines to calculate the:
#
# separation matrix "S(l,n,nue,x)" needed to transform
# the regular and outgoing wave functions if the coordinate
# system is shifted along the z-axis
#
# the matrix of rotation to transform the regular and outgoing wave
# functions if the coordinate system is rotated by use of the
# Eulerian angles of rotation
#
# the C21- and Q12 operators of the block operator for arbitrarily
# oriented bispheres (for both the rotation-shift, and the translation
# formulation)
#
# the T-matrices of sound soft (Dirichlet problem), hard
# (von Neumann problem), and penetrable (mixed problem) spheres
#
# the matrices "Y_l_J_u" and "Y_l_J_l" related to the Janus spheres.
# The elements are calculated from integrals over products of Legendre
# polynomials
#
# the diagonal matrices required for the Janus sphere
#
# the T-matrix for the s-p-Janus sphere
#
# the T-matrix for the h-p-Janus sphere
#
# the T-matrix for the h-s-Janus sphere
#
# the T-matrix for the sp-2layer sphere
#
# the T-matrix for the hp-2layer sphere
#

import numpy as np
import scipy.misc as scm
import scipy.special as scs
import scipy.integrate as sqa
```

```python
# Calculation of the separation matrix "S(l,n,nue,x)" according to
# [Martin, S. 104, Gl. (3.126)]:

def SM(el, en, nu, x):
    zi = 0 + 1j
    n1 = np.absolute(el)
    w = zi / 2 / x
    we = np.exp(zi * x) / zi / x
    pf = (-1)**n1 * zi**(en + nu) * we
    j_u = n1
    j_o = en + nu + 1
    S_sumj = 0.
    for j in range(j_u, j_o):
        du = [0, j - en]
        do = [nu, j - n1]
        s_u = max(du)
        s_o = min(do) + 1
        j1 = scs.factorial(j, exact=True)
        j2 = w**j
        S_sums = 0.
        for s in range(s_u, s_o):
            jms = j - s
            f1 = np.sqrt(2. * nu + 1)
            f1 = np.log(f1)
            f2 = np.sqrt(2. * en + 1)
            f2 = np.log(f2)
            f = f1 + f2
            a1 = 0.
            for i in range(2, nu + s + 1):
                a1 = a1 + np.log(float(i))
            a2 = 0.
            for i in range(2, en + jms + 1):
                a2 = a2 + np.log(float(i))
            a = a1 + a2
            b1 = 0.
            for i in range(2, s + 1):
                b1 = b1 + np.log(float(i))
            b1s = 0.
            for i in range(2, n1 + s + 1):
                b1s = b1s + np.log(float(i))
            b2 = 0.
            for i in range(2, nu - s + 1):
                b2 = b2 + np.log(float(i))
```

```python
        b3 = 0.
        for i in range(2, jms + 1):
            b3 = b3 + np.log(float(i))
        b3s = 0.
        for i in range(2, jms - n1 + 1):
            b3s = b3s + np.log(float(i))
        b4 = 0.
        for i in range(2, en - jms + 1):
            b4 = b4 + np.log(float(i))
        b = b1s + b1 + b2 + b3s + b3 + b4
        c1 = 0.
        for i in range(2, nu + n1 + 1):
            c1 = c1 + np.log(float(i))
        c2 = 0.
        for i in range(2, en - n1 + 1):
            c2 = c2 + np.log(float(i))
        c = c1 + c2
        d1 = 0.
        for i in range(2, nu - n1 + 1):
            d1 = d1 + np.log(float(i))
        d2 = 0.
        for i in range(2, en + n1 + 1):
            d2 = d2 + np.log(float(i))
        d = d1 + d2
        s_fak = np.exp(c - d)
        s_fak = np.sqrt(s_fak)
        s_pref = np.log(s_fak)
        res_fak = np.exp(a + f + s_pref - b)
        S_sums = S_sums + res_fak
      S_sumj = S_sumj + j1 * j2 * S_sums
  S_M = pf * S_sumj
  return S_M


# Calculation of the matrix of rotation "drm(n,l,l1,alpha,theta_p,gamma)"
# (note, that "alpha", "theta_p", and "gamma" must be given in degree!):

def drm(en, el, el_1, alpha, theta_p, gamma):
  alpha = alpha * np.pi / 180.
  gamma = gamma * np.pi / 180.
  zi = 0 + 1j
  n_p_l1 = en + el_1
  n_m_l1 = en - el_1
  n_p_l = en + el
```

```
n_m_l = en - el
l_p_l1 = el + el_1
s0 = [0, - l_p_l1]
s1 = [n_m_l, n_m_l1]
s_min = max(s0)
s_max = min(s1)
z1 = 0.
for i in range(2, n_p_l1 + 1):
    z1 = z1 + np.log(float(i))
z1 = z1 * 0.5
z2 = 0.
for i in range(2, n_m_l1 + 1):
    z2 = z2 + np.log(float(i))
z2 = z2 * 0.5
z3 = 0.
for i in range(2, n_p_l + 1):
    z3 = z3 + np.log(float(i))
z3 = z3 * 0.5
z4 = 0.
for i in range(2, n_m_l + 1):
    z4 = z4 + np.log(float(i))
z4 = z4 * 0.5
za = z1 + z2 + z3 + z4
D_m = 0.
for s in range(s_min, s_max + 1):
    n1 = 0.
    for i in range(2, s + 1):
        n1 = n1 + np.log(float(i))
    n2 = 0.
    for i in range(2, l_p_l1 + s + 1):
        n2 = n2 + np.log(float(i))
    n3 = 0.
    for i in range(2, n_m_l1 - s + 1):
        n3 = n3 + np.log(float(i))
    n4 = 0.
    for i in range(2, n_m_l - s + 1):
        n4 = n4 + np.log(float(i))
    na = n1 + n2 + n3 + n4
    d_m = np.exp(za - na)
    d_m = d_m * (-1.0)**(n_p_l1 + s)
    e_p0 = 2 * en - 2 * s - l_p_l1
    e_p1 = 2 * s + l_p_l1
    w_sin = (scs.sindg(theta_p / 2.)**e_p0 if e_p0 != 0 else 1.0)
```

```
      w_cos = scs.cosdg(theta_p / 2.)**e_p1
      d_m = d_m * w_sin * w_cos
      D_m = D_m + d_m
    D_m = np.exp(- zi * el * alpha) * D_m * np.exp(- zi * el_1 * gamma)
    return D_m


# Calculation of the C21- and Q12 operators of the block operator for
# arbitrarily oriented bispheres!


def Q12(n1cut, alpha, theta_p, sm_l_i_j, t_o, qr):
  c_l0_m0 = []
  for l0 in range(-n1cut, n1cut + 1):
    c_m0_3 = []
    for m0 in range(np.abs(l0), n1cut + 1):
      c1 = t_o[m0]
      c2 = 0.0
      for l1 in range(-n1cut, n1cut + 1):
        q_l1 = qr[l1 + n1cut]
        z_l1 = sm_l_i_j[np.abs(l1)]
        z_l1_m0 = z_l1[m0]
        D_rueck = drm(m0, l0, l1, alpha, theta_p, 0.)
        for n1 in range(np.abs(l1), n1cut + 1):
          n1z = n1 - np.abs(l1)
          S_rueck = z_l1_m0[n1]
          c2 = c2 + (-1)**m0 * q_l1[n1z] * S_rueck * D_rueck
      c2 = c2 * c1 * (-1)
      c_m0_3 = c_m0_3 + [c2]
    c_l0_m0 = c_l0_m0 + [c_m0_3]
  return c_l0_m0


def Q12_tl_it1(n1cut, alpha, theta_p, sm_l_i_j, t_o, qr):
  c_l0_m0 = []
  for l0 in range(-n1cut, n1cut + 1):
    c_m0_3 = []
    for m0 in range(np.abs(l0), n1cut + 1):
      c1 = t_o[m0]
      c2 = 0.0
      for l1 in range(-n1cut, n1cut + 1):
        q_l1 = qr[l1 + n1cut]
        z_l1 = sm_l_i_j[np.abs(l1)]
        z_l1_m0 = z_l1[m0]
        D_rueck = drm(m0, l0, l1, alpha, theta_p, 0.)
        for n1 in range(np.abs(l1), n1cut + 1):
```

```python
            D_hin = drm(n1, l1, 0, 0., -theta_p, -alpha)
            D_ges = D_hin * D_rueck
            n1z = n1 - np.abs(l1)
            S_rueck = z_l1_m0[n1]
            c2 = c2 + (-1)**m0 * q_l1[n1z] * S_rueck * D_ges
          c2 = c2 * c1 * (-1)
          c_m0_3 = c_m0_3 + [c2]
        c_l0_m0 = c_l0_m0 + [c_m0_3]
    return c_l0_m0


def C21(n1cut, alpha, theta_p, sm_l_i_j, t_o1, cr):
    q_l1_n1 = []
    for l1 in range(-n1cut, n1cut + 1):
        x_l1 = sm_l_i_j[np.abs(l1)]
        q_n1_3 = []
        for n1 in range(np.abs(l1), n1cut + 1):
            q1 = t_o1[n1]
            x_l1_n1 = x_l1[n1]
            q2 = 0.0
            for l0 in range(-n1cut, n1cut + 1):
                c_l0 = cr[l0 + n1cut]
                for m0 in range(np.abs(l0), n1cut + 1):
                    m0z = m0 - np.abs(l0)
                    S_rueck = x_l1_n1[m0]
                    D_rueck = drm(m0, l0, l1, alpha, theta_p, 0.)
                    D_rueck = np.conjugate(D_rueck)
                    q2 = q2 + (-1)**m0 * c_l0[m0z] * S_rueck * D_rueck
            q2 = q2 * q1 * (-1)
            q_n1_3 = q_n1_3 + [q2]
        q_l1_n1 = q_l1_n1 + [q_n1_3]
    return q_l1_n1


def C21_tl_it1(n1cut, alpha, theta_p, sm_l_i_j, t_o1, cr):
    q_l1_n1 = []
    for l1 in range(-n1cut, n1cut + 1):
        q_n1_3 = []
        for n1 in range(np.abs(l1), n1cut + 1):
            q1 = t_o1[n1]
            q2 = 0.0
            for l0 in range(-n1cut, n1cut + 1):
                c_l0 = cr[l0 + n1cut]
                x_l0 = sm_l_i_j[np.abs(l0)]
                x_l0_m0 = x_l0[n1]
```

```
        D_hin = drm(n1, l1, l0, alpha, theta_p, 0.)
        for m0 in range(np.abs(l0), n1cut + 1):
            m0z = m0 - np.abs(l0)
            S_rueck = x_l0_m0[m0]
            D_rueck = drm(m0, l0, 0, 0., -theta_p, -alpha)
            D_ges = D_hin * D_rueck
            q2 = q2 + (-1)**m0 * c_l0[m0z] * S_rueck * D_ges
        q2 = q2 * q1 * (-1)
        q_n1_3 = q_n1_3 + [q2]
    q_l1_n1 = q_l1_n1 + [q_n1_3]
  return q_l1_n1
```

Calculation of the T-matrix of a sound soft sphere:

```
def tm_s(ncut, beta):
  zi = 0. + 1.0j
  n = np.linspace(0,ncut,ncut+1,dtype=int)
  bes_1 = scs.spherical_jn(n,beta,derivative=False)
  neu_1 = scs.spherical_yn(n,beta,derivative=False)
  han_1 = bes_1 + zi * neu_1
  tm_soft = - bes_1 / han_1
  return tm_soft
```

Calculation of the T-matrix of a sound hard sphere:

```
def tm_h(ncut, beta):
  zi = 0. + 1.0j
  n = np.linspace(0,ncut,ncut+1,dtype=int)
  bes_1 = scs.spherical_jn(n,beta,derivative=True)
  neu_1 = scs.spherical_yn(n,beta,derivative=True)
  han_1 = bes_1 + zi * neu_1
  tm_hard = - bes_1 / han_1
  return tm_hard
```

Calculation of the T-matrix of a sound penetrable sphere:

```
def tm_p(ncut, beta, a, k, k_p_k, rho_p):
  zi = 0. + 1.0j
  beta_p = k * k_p_k * a
  kap_k = k_p_k
  kap_r = rho_p
  n = np.linspace(0,ncut,ncut+1,dtype=int)
  bes = scs.spherical_jn(n,beta)
```

```python
    bes_1 = scs.spherical_jn(n,beta,derivative=True)
    bes_p = scs.spherical_jn(n,beta_p)
    bes_p_1 = scs.spherical_jn(n,beta_p,derivative=True)
    neu = scs.spherical_yn(n,beta)
    neu_1 = scs.spherical_yn(n,beta,derivative=True)
    han = bes + zi * neu
    han_1 = bes_1 + zi * neu_1
    tz = kap_k / kap_r * bes * bes_p_1 - bes_1 * bes_p
    tn = kap_r / kap_k * han_1 * bes_p - bes_p_1 * han
    tm_pene = kap_r / kap_k * tz / tn
    return tm_pene


# Calculation of the matrices of Janus spheres:


def a_le_p(l, n, x):
    alp_ln = scs.lpmn(n, n, np.cos(x))
    alep = alp_ln[0]
    alep = alep[l]
    return alep


def Y_l_J_u(l, ncut, theta_j):
    al = np.abs(l)
    theta_j = theta_j * np.pi / 180.
    Y_l_J_n_m = []
    for n in range(al, ncut + 1):
        Y_l_J_m = []
        for m in range(al, ncut + 1):
            mf = scm.factorial((m - al), exact=True) / \
                scm.factorial((m + al), exact=True)
            nf = scm.factorial((n - al), exact=True) / \
                scm.factorial((n + al), exact=True)
            c_n_m = np.sqrt((2. *n + 1) * (2. * m + 1) * mf * nf) / 2.
            y_n_m = lambda x: np.sin(x) * a_le_p(al, ncut, x)[n] * \
                a_le_p(al, ncut, x)[m]
            Y_l_J_m = Y_l_J_m + [c_n_m * sqa.quad(y_n_m, theta_j, np.pi, \
                        epsabs=1.0e-06, epsrel=1.0e-06)[0]]
        Y_l_J_n_m = Y_l_J_n_m + [Y_l_J_m]
    P_l_Ju_n_m = np.array(Y_l_J_n_m)
    return P_l_Ju_n_m


def Y_l_J_l(l, ncut, theta_j):
    al = np.abs(l)
    theta_j = theta_j * np.pi / 180.
```

```python
    Y_l_J_n_m = []
    for n in range(al, ncut + 1):
        Y_l_J_m = []
        for m in range(al, ncut + 1):
            mf = scm.factorial((m - al), exact=True) / \
                scm.factorial((m + al), exact=True)
            nf = scm.factorial((n - al), exact=True) / \
                scm.factorial((n + al), exact=True)
            c_n_m = np.sqrt((2. *n + 1) * (2. * m + 1) * mf * nf) / 2.
            y_n_m = lambda x: np.sin(x) * a_le_p(al, ncut, x)[n] * \
                a_le_p(al, ncut, x)[m]
            Y_l_J_m = Y_l_J_m + [c_n_m * sqa.quad(y_n_m, 0.0, theta_j, \
                epsabs=1.0e-06, epsrel=1.0e-06)[0]]
        Y_l_J_n_m = Y_l_J_n_m + [Y_l_J_m]
    P_l_Jl_n_m = np.array(Y_l_J_n_m)
    return P_l_Jl_n_m


def diag_h(l, beta, ncut):
    al = np.abs(l)
    zi = 0. + 1.0j
    D_h = []
    for i in range(al, ncut + 1):
        bes = scs.spherical_jn(i,beta)
        neu = scs.spherical_yn(i,beta)
        han = bes + zi * neu
        D = []
        for j in range(al, ncut + 1):
            diag = (han if j == i else 0.0)
            D = D + [diag]
        D_h = D_h + [D]
    D_h = np.matrix(D_h)
    return D_h


def diag_j(l, beta, ncut):
    al = np.abs(l)
    D_j = []
    for i in range(al, ncut + 1):
        bes = scs.spherical_jn(i,beta)
        D = []
        for j in range(al, ncut + 1):
            diag = (bes if j == i else 0.0)
            D = D + [diag]
        D_j = D_j + [D]
```

```python
    D_j = np.matrix(D_j)
    return D_j


def diag_jp(l, beta_p, ncut):
    al = np.abs(l)
    D_jp = []
    for i in range(al, ncut + 1):
        bes = scs.spherical_jn(i,beta_p)
        D = []
        for j in range(al, ncut + 1):
            diag = (bes if j == i else 0.0)
            D = D + [diag]
        D_jp = D_jp + [D]
    D_jp = np.matrix(D_jp)
    return D_jp


def diag_hs(l, beta, ncut):
    al = np.abs(l)
    zi = 0. + 1.0j
    D_hs = []
    for i in range(al, ncut + 1):
        bes = scs.spherical_jn(i,beta,derivative=True)
        neu = scs.spherical_yn(i,beta,derivative=True)
        han = bes + zi * neu
        D = []
        for j in range(al, ncut + 1):
            diag = (han if j == i else 0.0)
            D = D + [diag]
        D_hs = D_hs + [D]
    D_hs = np.matrix(D_hs)
    return D_hs


def diag_js(l, beta, ncut):
    al = np.abs(l)
    D_js = []
    for i in range(al, ncut + 1):
        bes = scs.spherical_jn(i,beta,derivative=True)
        D = []
        for j in range(al, ncut + 1):
            diag = (bes if j == i else 0.0)
            D = D + [diag]
        D_js = D_js + [D]
    D_js = np.matrix(D_js)
```

```python
      return D_js

def diag_jps(l, k_p_k, rho_p, beta_p, ncut):
   al = np.abs(l)
   kappa = rho_p / k_p_k
   D_jps = []
   for i in range(al, ncut + 1):
      bes = scs.spherical_jn(i,beta_p,derivative=True) / kappa
      D = []
      for j in range(al, ncut + 1):
         diag = (bes if j == i else 0.0)
         D = D + [diag]
      D_jps = D_jps + [D]
   D_jps = np.matrix(D_jps)
   return D_jps

def e_matrix(l, ncut):
   al = np.abs(l)
   c11 = []
   for i in range(al, ncut + 1):
      c1 = []
      for j in range(al, ncut + 1):
         c = (1.0 if j == i else 0.0)
         c1 = c1 + [c]
      c11 = c11 + [c1]
   E = np.matrix(c11)
   return E

# Calculation of the T-matrix for the h-s-Janus sphere:

def t_hs(l, beta, theta_j, ncut):
   m_h = diag_h(l, beta, ncut)
   m_hs = diag_hs(l, beta, ncut)
   m_j = diag_j(l, beta, ncut)
   m_js = diag_js(l, beta, ncut)
   pa_l_j = Y_l_J_l(l, ncut, theta_j)
   pb_l_j = Y_l_J_u(l, ncut, theta_j)

   m_h_p = pb_l_j * m_h
   m_hs_p = pa_l_j * m_hs
   m_j_p = pb_l_j * m_j
   m_js_p = pa_l_j * m_js
   A = m_hs_p + m_h_p
```

```python
    B = m_js_p + m_j_p
    tm_hs = - np.linalg.inv(A) * B
    return tm_hs
```

```python
# Calculation of the T-matrix for the s-p-Janus sphere:

def t_sp(l, beta, beta_p, theta_j, k_p_k, rho_p, ncut):
    m_h = diag_h(l, beta, ncut)
    m_j = diag_j(l, beta, ncut)
    m_hs = diag_hs(l, beta, ncut)
    m_js = diag_js(l, beta, ncut)
    pu_l_j = Y_l_J_u(l, ncut, theta_j)
    pl_l_j = Y_l_J_l(l, ncut, theta_j)
    d_jp = diag_jp(l, beta_p, ncut)
    d_jps = diag_jps(l, k_p_k, rho_p, beta_p, ncut)

    m_dcs = m_hs - k_p_k / rho_p * m_h * d_jps * np.linalg.inv(d_jp)
    m_dds = m_js - k_p_k / rho_p * m_j * d_jps * np.linalg.inv(d_jp)
    A = pl_l_j * m_h + pu_l_j * m_dcs
    B = pl_l_j * m_j + pu_l_j * m_dds
    tm_sp = - np.linalg.inv(A) * B
    return tm_sp
```

```python
# Calculation of the T-matrix for the h-p-Janus sphere:

def t_hp(l, beta, beta_p, theta_j, k_p_k, rho_p, ncut):
    m_h = diag_h(l, beta, ncut)
    m_j = diag_j(l, beta, ncut)
    m_hs = diag_hs(l, beta, ncut)
    m_js = diag_js(l, beta, ncut)
    pu_l_j = Y_l_J_u(l, ncut, theta_j)
    pl_l_j = Y_l_J_l(l, ncut, theta_j)
    d_jp = diag_jp(l, beta_p, ncut)
    d_jps = diag_jps(l, k_p_k, rho_p, beta_p, ncut)

    m_dcs = m_h - rho_p * m_hs * d_jp * np.linalg.inv(d_jps) / k_p_k
    m_dds = m_j - rho_p * m_js * d_jp * np.linalg.inv(d_jps) / k_p_k
    A = pl_l_j * m_hs + pu_l_j * m_dcs
    B = pl_l_j * m_js + pu_l_j * m_dds
    tm_hp = - np.linalg.inv(A) * B
    return tm_hp
```

Calculation of the T-matrix of the sp-2-layer sphere:

```python
def tm_sp(ncut, beta_b, a, b, k, k_p_k, rho_p):
    zi = 0. + 1.0j
    beta_pa = k * k_p_k * a
    ts = tm_s(ncut, beta_pa)
    beta_pb = k * k_p_k * b
    kap_k = k_p_k
    kap_r = rho_p
    n = np.linspace(0,ncut,ncut+1,dtype=int)
    bes = scs.spherical_jn(n,beta_b)
    bes_1 = scs.spherical_jn(n,beta_b,derivative=True)
    bes_b = scs.spherical_jn(n,beta_pb)
    bes_1_b = scs.spherical_jn(n,beta_pb,derivative=True)
    neu = scs.spherical_yn(n,beta_b)
    neu_1 = scs.spherical_yn(n,beta_b,derivative=True)
    han = bes + zi * neu
    han_1 = bes_1 + zi * neu_1
    neu_b = scs.spherical_yn(n,beta_pb)
    neu_1_b = scs.spherical_yn(n,beta_pb,derivative=True)
    han_b = bes_b + zi * neu_b
    han_1_b = bes_1_b + zi * neu_1_b
    bes_p = scs.spherical_jn(n,beta_pb) + ts[n] * han_b
    bes_p_1 = scs.spherical_jn(n,beta_pb,derivative=True) + ts[n] * han_1_b
    tz = kap_k / kap_r * bes * bes_p_1 - bes_1 * bes_p
    tn = kap_r / kap_k * han_1 * bes_p - bes_p_1 * han
    tm_2l_sp = kap_r / kap_k * tz / tn
    return tm_2l_sp
```

Calculation of the T-matrix of the hp-2-layer sphere:

```python
def tm_hp(ncut, beta_b, a, b, k, k_p_k, rho_p):
    zi = 0. + 1.0j
    beta_pa = k * k_p_k * a
    th = tm_h(ncut, beta_pa)
    beta_pb = k * k_p_k * b
    kap_k = k_p_k
    kap_r = rho_p
    n = np.linspace(0,ncut,ncut+1,dtype=int)
    bes = scs.spherical_jn(n,beta_b)
    bes_1 = scs.spherical_jn(n,beta_b,derivative=True)
    bes_b = scs.spherical_jn(n,beta_pb)
    bes_1_b = scs.spherical_jn(n,beta_pb,derivative=True)
```

```
neu = scs.spherical_yn(n,beta_b)
neu_1 = scs.spherical_yn(n,beta_b,derivative=True)
han = bes + zi * neu
han_1 = bes_1 + zi * neu_1
neu_b = scs.spherical_yn(n,beta_pb)
neu_1_b = scs.spherical_yn(n,beta_pb,derivative=True)
han_b = bes_b + zi * neu_b
han_1_b = bes_1_b + zi * neu_1_b
bes_p = scs.spherical_jn(n,beta_pb) + th[n] * han_b
bes_p_1 = scs.spherical_jn(n,beta_pb,derivative=True) + th[n] * han_1_b
tz = kap_k / kap_r * bes * bes_p_1 - bes_1 * bes_p
tn = kap_r / kap_k * han_1 * bes_p - bes_p_1 * han
tm_2l_hp = kap_r / kap_k * tz / tn
return tm_2l_hp
```

```python
#        Program "plot_Bessel_function"
#
# This program plots the spherical Bessel function for real-valued arguments
#  z in the interval [0.,15.]. The results can be compared to Fig. 10.1
#  in reference [6].
#

print()
print()
print("        --- Plot of spherical Bessel functions! ---")
print()
print()

import matplotlib.pyplot as plt
import numpy as np
import scipy.special as scs

# Input order of Bessel function:

n = int(input('Order of Bessel function ... n: '))

# Calculation of the Bessel function of order "n" in [0,14]:

z = np.linspace(0.0, 15.0, 151)
bes = scs.spherical_jn(n,z,derivative=False)

# Plot of Bessel function:

#plt.yscale('log')        # use this command for lin-log plot!
plt.plot(z, bes, color = 'black', linewidth=2.0)
plt.xlabel("z", fontsize=16)
plt.ylabel("spherical Bessel function j_n(z)", fontsize=16)
plt.show()
```

```
#           Program "trans_single_wf_co.py"
#
# Transformation of the regular- and radiating wave functions "psi_{l=0,n}"
# and "phi_{l=0,n}" for a dimensionless shift "k0b" along the positive z-axis
# of the laboratory frame. The original and shifted wave functions are
# calculated at different angles "theta" in [0, \pi] in steps of
# "180/w_num_m" degrees for a fixed but dimensionless distance "k_0r" in
# the laboratory system!
#

print()
print()
print(" --- Translation of a regular- and radiating  wave ...")
print("     ... function along the positive z-axis! --- ")

print()
print()

import numpy as np
import scipy as scp
import scipy.special as scs
import basics as bas
import matplotlib.pyplot as plt

# Input order of Legendre polynomial, the transformation parameters,
# and the truncatuion parameter:

en = int(input('order of Legendre polynomial ... n: '))
k0r = float(input('dimensionless distance in laboratory frame ... k0r: '))
k0b = float(input('dimensionless shift along z-axis ... k0b: '))
nucut = int(input('truncation parameter ... nucut: '))

w_num = 37     # number of angles "\theta" in [0°,180°]!
w_num_m = w_num - 1
zi = 0. + 1.0j
theta = np.linspace(0.0, 180.0, w_num)
ctheta = np.cos(theta * scp.pi / 180.)

# Calculation of the dimensionless distances "kr1" and angles "theta1" in
# the shifted c.-system:

theta1 = []
ctheta1 = []
```

```python
kOr1 = []
for i in range(0,w_num):
    kOr1 = kOr1 + [np.sqrt(kOb**2 + kOr**2 - 2. * kOb * kOr * ctheta[i])]
    if i == 0:
        arg_cos = -1.0
    elif i == w_num_m:
        arg_cos = 1.0
    else:
        arg_cos = -(kOr**2 - kOr1[i]**2 - kOb**2) / (2. * kOr1[i] * kOb)
    theta1 = theta1 + [180. * ( 1. - np.arccos(arg_cos) / scp.pi)]
    if kOb <= kOr:
        theta1[0] = 0.0
    else:
        theta1[0] = 180.0
    ctheta1 = ctheta1 + [np.cos(theta1[i] * scp.pi / 180.)]
ctheta1 = np.array(ctheta1)
theta1 = np.array(theta1)
kr1 = np.array(kOr1)

# Calculation of the separation matrix:

y_os = []
for s_sum in range(0, nucut + 1):
    y_os = y_os + [bas.SM(0, en, s_sum, kOb)]
y_o1 = np.real(y_os)

# calculation of the wave functions in the laboratory frame:

psi_n = []
psi_n_trans = []
phi_n = []
phi_n_trans = []
for j in range(0, w_num):
    pn0 = scs.lpn(en,ctheta[j])
    pn1 = pn0[0]
    ysh = np.sqrt((2. * en + 1) / 4. / np.pi) * float(pn1[en])
    u = scs.spherical_jn(en,kOr)
    v0 = scs.spherical_yn(en,kOr)
    v = u + zi * v0
    psi_n = psi_n + [u * ysh]
    phi_n = phi_n + [v * ysh]
```

transformation of the wave functions of the laboratory frame:

```python
    pn0_o1 = scs.lpn(nucut,ctheta1[j])
    pn1_o1 = pn0_o1[0]
    tpsi = 0.
    tphi = 0.
    for nu in range(0, nucut + 1):
        ysh_o1 = np.sqrt((2. * nu + 1) / 4. / np.pi) * float(pn1_o1[nu])
        u_o1 = scs.spherical_jn(nu,kr1[j])
        v0_o1 = scs.spherical_yn(nu,kr1[j])
        v_o1 = u_o1 + zi * v0_o1
        psi_nu_xr1 = u_o1 * ysh_o1
        phi_nu_xr1 = v_o1 * ysh_o1
        tpsi = tpsi + y_o1[nu] * psi_nu_xr1
        if kr1[j] < k0b:
            tphi = tphi + y_os[nu] * psi_nu_xr1
        else:
            tphi = tphi + y_o1[nu] * phi_nu_xr1
    tpsi = (-1)**en * tpsi
    tphi = (-1)**en * tphi
    psi_n_trans = psi_n_trans + [tpsi]
    phi_n_trans = phi_n_trans + [tphi]

psi_n = np.array(psi_n)
psi_n_trans = np.array(psi_n_trans)
phi_n = np.array(phi_n)
phi_n_trans = np.array(phi_n_trans)

# Plot of the results:

plt.plot(1)
plt.plot(theta, psi_n, 'o', color = 'black', linewidth = 2.0, label = \
    "\psi_n")
plt.plot(theta, psi_n_trans, color = 'black', linewidth = 2.0, label = \
    "\psi_n_trans")
plt.legend()
plt.xlabel("$\\theta$ (degree)", fontsize=16)
plt.show()

plt.plot(2)
plt.plot(theta, np.real(phi_n), 'o', color = 'black', linewidth = 2.0, label = \
    "real(\phi_n)")
plt.plot(theta, np.real(phi_n_trans), color = 'black', linewidth = 2.0, \
```

```
        label = "real(\phi_n_trans)")
plt.legend()
plt.xlabel("$\\theta$ (degree)", fontsize=16)
plt.show()


plt.plot(3)
plt.plot(theta, np.imag(phi_n), 'o', color = 'black', linewidth = 2.0, label = \
        "imag(\phi_n)")
plt.plot(theta, np.imag(phi_n_trans), color = 'black', linewidth = 2.0, \
        label = "imag(\phi_n_trans)")
# plt.legend()
plt.legend(loc = 'lower right')
plt.xlabel("$\\theta$ (degree)", fontsize=16)
plt.show()
```

```
#            Program "trans_single_wf_co1.py"
#
# Transformation of the regular- and radiating wave functions "psi_{l=0,n}"
# and "phi_{l=0,n}" for a dimensionless shift "k0b" along the positive z-axis
# of the laboratory frame. The original and shifted wave functions are
# calculated at different angles "theta1" in [0, \pi] in steps of
# "180/w_num_m" degrees, and for a fixed but dimensionless distance "k_0r_1" in
# the shifted co.-system!
#


print()
print()
print(" --- Translation of a regular- and radiating  wave function ...")
print("     ... along the positive z-axis. k0b >= 2 * k0r1 ! --- ")
print()
print()


import numpy as np
import scipy as scp
import scipy.special as scs
import basics as bas
import matplotlib.pyplot as plt


# Input order of Legendre polynomial, the transformation parameters,
# and the truncation paerameter:

en = int(input('order of Legendre polynomial  n: '))
k0r1 = float(input('dimensionless distance in the shifted system ... k0r1: '))
a1 = 2. * k0r1
a1 = np.str(a1)
a0 = "dimensionless shift along positive z-axis (k0b >= "
a2 = "!)  k0b: "
a = a0 + a1 + a2
k0b = float(input(a))
nucut = int(input('truncation parameter ... nucut: '))
print()


w_num = 37       # number of angles "\theta1" in [0°,180°]!
w_num_m = w_num - 1
zi = 0. + 1.0j
theta1_o1 = np.linspace(0.0, 180.0, w_num)
ctheta1_o1 = np.cos(theta1_o1 * scp.pi / 180.)
xa = np.arange(w_num)
```

```python
xachse = np.arange(w_num)

# calculation of the separation matrix:

y_os = []
for s_sum in range(0, nucut + 1):
    y_os = y_os + [bas.SM(0, en, s_sum, k0b)]
y_o1 = np.real(y_os)

# calculation of the dimensionless distances "kr" and angles "theta" in
# the laboratory frame:

theta1 = 180. - theta1_o1
ctheta1 = np.cos(theta1 * scp.pi / 180.)
theta = []
ctheta = []
k0r = []
for i in range(0,w_num):
    k0r = k0r + [np.sqrt(k0b**2 + k0r1**2 - 2. * k0b * k0r1 * ctheta1[i])]
    if i == 0 or i == w_num_m:
        arg_cos = 1.0
    else:
        arg_cos = -(k0r1**2 - k0r[i]**2 - k0b**2) / (2. * k0r[i] * k0b)
    theta = theta + [180.0 * np.arccos(arg_cos) / scp.pi]
    ctheta = ctheta + [np.cos(theta[i] * scp.pi / 180.)]
ctheta = np.array(ctheta)
theta = np.array(theta)
kr = np.array(k0r)

# calculation of the wave functions in the laboratory frame:

psi_n = []
psi_n_trans = []
phi_n = []
phi_n_trans = []
for j in range(0, w_num):
    pn0 = scs.lpn(en,ctheta[j])
    pn1 = pn0[0]
    ysh = np.sqrt((2. * en + 1) / 4. / np.pi) * float(pn1[en])
    u = scs.spherical_jn(en,kr[j])
    v0 = scs.spherical_yn(en,kr[j])
    v = u + zi * v0
    psi_n = psi_n + [u * ysh]
```

```
    phi_n = phi_n + [v * ysh]

# transformation of the wave functions of the laboratory frame:

    pn0_o1 = scs.lpn(nucut,ctheta1_o1[j])
    pn1_o1 = pn0_o1[0]
    tpsi = 0.
    tphi = 0.
    for nu in range(0, nucut + 1):
        ysh_o1 = np.sqrt((2. * nu + 1) / 4. / np.pi) * pn1_o1[nu]
        u_o1 = scs.spherical_jn(nu,k0r1)
        v0_o1 = scs.spherical_yn(nu,k0r1)
        v_o1 = u_o1 + zi * v0_o1
        psi_nu_xr1 = u_o1 * ysh_o1
        phi_nu_xr1 = v_o1 * ysh_o1
        tpsi = tpsi + y_o1[nu] * psi_nu_xr1
        tphi = tphi + y_os[nu] * psi_nu_xr1
    tpsi = (-1)**en * tpsi
    tphi = (-1)**en * tphi
    psi_n_trans = psi_n_trans + [tpsi]
    phi_n_trans = phi_n_trans + [tphi]

psi_n = np.array(psi_n)
psi_n_trans = np.array(psi_n_trans)
phi_n = np.array(phi_n)
phi_n_trans = np.array(phi_n_trans)

# Plot of the results:

plt.plot(1)
plt.plot(theta1_o1, psi_n, 'o', color = 'black', linewidth = 2.0, label = \
        "\psi_n")
plt.plot(theta1_o1, psi_n_trans, color = 'black', linewidth = 2.0, label = \
        "\psi_n_trans")
plt.legend()
plt.xlabel("$\\theta_1$ (degree)", fontsize=16)
plt.show()

plt.plot(2)
plt.plot(theta1_o1, np.real(phi_n), 'o', color = 'black', linewidth = 2.0, label = \
        "real(\phi_n)")
plt.plot(theta1_o1, np.real(phi_n_trans), color = 'black', linewidth = 2.0, \
        label = "real(\phi_n_trans)")
```

```
plt.legend()
plt.xlabel("$\\theta_1$ (degree)", fontsize=16)
plt.show()

plt.plot(3)
plt.plot(theta1_o1, np.imag(phi_n), 'o', color = 'black', linewidth = 2.0, label = \
    "imag(\phi_n)")
plt.plot(theta1_o1, np.imag(phi_n_trans), color = 'black', linewidth = 2.0, \
    label = "imag(\phi_n_trans)")
plt.legend()
plt.xlabel("$\\theta_1$ (degree)", fontsize=16)
plt.show()
```

```
#        Program "sumtest.py"
#
# This program provides a test of (1.80) of the manuscript.
#

print()
print()
print("              --- Sum test! ---")
print()
print()

import numpy as np
import basics as bas

k0b = float(input(' real ................... k0b : '))
n1 = int(input(' integer ................. n1 : '))
ncut = int(input(' truncation parameter .... ncut : '))
print()
print()

zi = 0. + 1.0j

z_ana = np.exp(zi * k0b) * np.sqrt(4. * np.pi * (2. * n1 + 1.))
z_sum = 0.
for n0 in range(0, ncut + 1):
    a = (- zi)**(n0 + n1)
    b = np.sqrt(4. * np.pi * (2. * n0 + 1.))
    c = bas.SM(0, n0, n1, k0b)
    c = np.real(c)
    z_sum = z_sum + a * b * c
print(' analytical value:  ', z_ana)
print(' approximation:     ', z_sum)
print()
print()
print(' Press Enter to finish')
input()
```

```
#        Program "plane_wave_trans.py"
#
# Comparison of the analytical expression and the expansion of the primary
# incident plane wave after translation from the laboratory frame into the
# body frame (rotation - shift - rotation). Calculation is performed along
# the surface of a shifted sphere with radius "a1", and for "theta_1" in
# [0, pi] (in steps of 10 degree)!
#


print()
print()
print("      Translation of the primary incident plane wave!")
print()
print()


import numpy as np
import basics as bas
import scipy as scp
import scipy.special as scs

# Input parameters:

a1 = float(input(' radius of the shifted sphere ... a1 [mm]: '))
b = float(input(' shift along the new z-direction after 1. rotation ... b [mm]: '))
alpha = float(input(' Eulerian angle ... alpha [deg.]: '))
theta_p = float(input(' Eulerian angle ... theta_p [deg.]: '))
ncut = int(input(" truncation parameter ...  ncut: "))
print()
print()


k0 = 1.0
k0b = k0 * b
beta_1 = k0 * a1
zi = 0. + 1.0j
pf = - zi / k0

theta_l = 180.0
theta_l1 = 19
theta = np.linspace(0.0, theta_l, theta_l1)
r_theta = theta * scp.pi / 180.
r_thetap = theta_p * scp.pi / 180.
```

```
# Calculation of the separation matrix:

sm_l_i_j = []
for l in range(0, ncut + 1):
    sm_i_j = []
    for i in range(0, ncut + 1):
        sm_j = []
        for j in range(0, ncut + 1):
            sm_j = sm_j + [bas.SM(l, i, j, k0b)]
        sm_i_j = sm_i_j + [sm_j]
    sm_l_i_j = sm_l_i_j + [sm_i_j]
sm_l_i_j_real = np.real(sm_l_i_j)


# Calculation of the primary incident plane wave in the body frame:

for i in range(0, theta_l1):
    sum_zw = 0.0
    ana_zw = np.exp(zi * k0b * np.cos(r_thetap) + zi * beta_1 *\
            np.cos(r_theta[i]))
    for n_zs in range(0, ncut + 1):
        bes = scs.spherical_jn(n_zs,beta_1)
        for l1 in range(-n_zs, n_zs + 1):
            Y_sh = scs.sph_harm(l1,n_zs,0.,r_theta[i])
            for ls in range(-ncut, ncut + 1):
                s_ls = sm_l_i_j_real[np.abs(ls)]
                s_ls_nzs = s_ls[n_zs]
                for n in range(np.abs(ls), ncut + 1):
                    d_n = np.sqrt(4. * np.pi * (2 * n + 1)) * zi**n
                    d_n = (-1)**n * d_n
                    Dm_1 = bas.drm(n, ls, 0, 0., -theta_p, -alpha)
                    Dm_2 = bas.drm(n_zs, l1, ls, alpha, theta_p, 0.)
                    S_m = s_ls_nzs[n]
                    sum_zw = sum_zw + d_n * Dm_1 * Dm_2 * S_m * bes * Y_sh
    print(' real ana vs. sum:')
    print(' theta = ',theta[i],'deg.: ',np.real(ana_zw),np.real(sum_zw))
    print(' imag ana vs. sum')
    print(' theta = ',theta[i],'deg.: ',np.imag(ana_zw),np.imag(sum_zw))
    print()
print(' Press Enter to finish')
input()
```

Appendix B
Full Python Programs of Chap. 2

© Springer Nature Switzerland AG 2020
T. Rother, *Sound Scattering on Spherical Objects*,
https://doi.org/10.1007/978-3-030-36448-9

```
#          Program "centered_sphere_input.py"
#

print()
print('    --- Input scattering configuration for acoustic plane ...')
print('         ... wave scattering on a centered sphere! ---')
print()

import os

os.system("del input_data_centered_sphere.txt")
os.system("type nul > input_data_centered_sphere.txt")
fobj = open("input_data_centered_sphere.txt", "w")
para = str(input('soft (s), hard (h), or penetrable (p) sphere? '))
print()
if para == 'p':
    k_p_k = float(input('ratio of wave numbers k_p/k_0: '))
    rho_p = float(input('density of the material inside the sphere rho_p: '))
else:
    k_p_k = 1.0
    rho_p = 1.0
a = float(input('radius a [mm]: '))
beta = float(input('size parameter beta: '))
print()
plot_para = str(input('lin-lin (ll), or lin-log (lg) plot? '))
s0 = "para = " + para + "\n"
fobj.write(s0)
s1 = str(k_p_k)
s1 = "k_p_k = " + s1 + "\n"
fobj.write(s1)
s2 = str(rho_p)
s2 = "rho_p = " + s2 + "\n"
fobj.write(s2)
s3 = str(a)
s3 = "a = " + s3 + "\n"
fobj.write(s3)
s4 = str(beta)
s4 = "beta = " + s4 + "\n"
fobj.write(s4)
s5 = "plot_para = " + plot_para
fobj.write(s5)
fobj.close()
```

```
#          Program "centered_sphere.py"
#
# This program calculates the plane wave scattering behaviour of a sound
# soft, hard, or penetrable sphere (homogeneous Dirichlet-, von
# Neumann- or mixed condition) centered in the laboratory frame.
# Note that the density of the material "rho_0" outside the spheres is
# always given by 1, and that the truncation parameter "ncut" is fixed
# to "ncut = beta + 5".
#

import scipy as scp
import scipy.special as scs
import matplotlib.pyplot as plt
import basics as bas
import os

# Reading scattering configuration from input file:

fobj = open("input_data_centered_sphere.txt", "r")

z = []
for line in fobj:
    line = line.strip()
    arr = line.split("= ")
    wert = str(arr[1])
    z = z + [wert]
fobj.close()

para = z[0]
k_p_k = float(z[1])
rho_p = float(z[2])
a = float(z[3])
beta = float(z[4])
plot_para = z[5]

# Fixing the truncation parameter "ncut" and the wave number "k0":

ncut = int(beta + 5.) # must possibly be increased for higher size parameters
k0 = beta / a

# Calculation of the T-matrix:

if para == 's':
```

```python
        tm = bas.tm_s(ncut, beta)
    elif para == 'h':
        tm = bas.tm_h(ncut, beta)
    else:
        tm = bas.tm_p(ncut, beta, a, k0, k_p_k, rho_p)


# Calculation of the differential scattering cross-section "dscross"
# as a function of the scattering angle "theta" in the interval [0, \pi]
# in steps of 0.5 degree:

theta = scp.linspace(0.0, 180.0, 361)
ctheta = scp.cos(theta * scp.pi / 180.)
zi = 0. + 1.0j
pref = zi / k0
dscross = []
psi_s = []
for i in range(0, 361):
    psi = 0.0
    y = scs.lpn(ncut,ctheta[i])
    y1 = y[0]
    for n in range(0, ncut +1):
        d_n = (2 * n + 1)
        sum_n = - d_n * y1[n] * tm[n]
        psi = psi + sum_n
    psi = pref * psi
    psi_s = psi_s + [psi]
    dscross = dscross + [psi * scp.conj(psi)]
dscross = scp.real(dscross)


# Generation of the result file:

os.system("del dscross_centered_sphere.txt")
os.system("type nul > dscross_centered_sphere.txt")
fobj1 = open("dscross_centered_sphere.txt", "w")
nl = "\n"
for i in range(0, 361):
    a1 = str(theta[i])
    a2 = " = "
    a3 = str(dscross[i]) + nl
    a = a1 + a2 + a3
    fobj1.write(a)
fobj1.close()
```

```
# Calculation of the total scattering cross-section "scat_tot" by use
# of the optical theorem:

print()
print()
print('Results: ')
w = scp.imag(psi_s[0])
scat_tot = 4 * scp.pi * w / k0
print()
print("total scattering cross-section: scat_tot = ", scat_tot)

# Plot of "dscross":

if plot_para == 'lg':
    plt.yscale('log')
    plt.plot(theta, dscross, color = 'black', linewidth=2.0)
else:
    plt.plot(theta, dscross, color = 'black', linewidth=2.0)
plt.xlabel("scattering angle (degree)", fontsize=16)
plt.ylabel("diff. scat. cross-sect.", fontsize=16)
plt.show()
```

```python
#          Program "centered_2l_sphere_input.py"
#

print()
print(' --- Input scattering configuration for acoustic plane ...')
print('     ... wave scattering on a centered 2layer sphere! ---')
print()

import os
os.system("del input_data_centered_2l_sphere.txt")
os.system("type nul > input_data_centered_2l_sphere.txt")
fobj = open("input_data_centered_2l_sphere.txt", "w")

# Input scattering configuration:

para = str(input('soft+penetrable (sp) or hard+penetrable (hp)? '))
print()
k_p_k = float(input('ratio of wave numbers k_p/k0: '))
rho_p = float(input('density of the material inside the sphere ... rho_p: '))
a = float(input('radius of the core ... a [mm]: '))
ts = float(input('thickness of the shell ... ts [mm]: '))
beta_b = float(input('size parameter of the 2l-sphere ... beta_b: '))
print()
ncut = int(input('truncation parameter  ... ncut: '))
print()
plot_para = str(input('lin-lin (ll), or lin-log (lg) plot? '))

s0 = "para = " + para + "\n"
fobj.write(s0)
s1 = str(k_p_k)
s1 = "k_p_k = " + s1 + "\n"
fobj.write(s1)
s2 = str(rho_p)
s2 = "rho_p = " + s2 + "\n"
fobj.write(s2)
s3 = str(a)
s3 = "a = " + s3 + "\n"
fobj.write(s3)
s4 = str(ts)
s4 = "ts = " + s4 + "\n"
fobj.write(s4)
s5 = str(beta_b)
s5 = "beta_b = " + s5 + "\n"
```

```
fobj.write(s5)
s6 = str(ncut)
s6 = "ncut = " + s6 + "\n"
fobj.write(s6)
s7 = "plot_para = " + plot_para
fobj.write(s7)
fobj.close()
```

```
#          Program "centered_2l_sphere.py"
#
# This program calculates the plane wave scattering behaviour of a
# two-layer sphere with a sound soft or hard core and a sound
# penetrable shell. Note, that the density of the material
# "rho_0" outside the spheres is always given by 1.
#

import scipy as scp
import scipy.special as scs
import matplotlib.pyplot as plt
import basics as bas
import os

# Reading scattering configuration from input file:

fobj = open("input_data_centered_2l_sphere.txt", "r")

z = []
for line in fobj:
    line = line.strip()
    arr = line.split("= ")
    wert = str(arr[1])
    z = z + [wert]
fobj.close()

para = z[0]
k_p_k = float(z[1])
rho_p = float(z[2])
a = float(z[3])
ts = float(z[4])
beta_b = float(z[5])
ncut = int(z[6])
plot_para = z[7]

# wave number "k" (outer space) and total radius of the 2-layer sphere:

b = a + ts
k0 = beta_b / b

# Calculation of the T-matrix:

if para == 'sp':
```

```
  tm = bas.tm_sp(ncut, beta_b, a, b, k0, k_p_k, rho_p)
else:
  tm = bas.tm_hp(ncut, beta_b, a, b, k0, k_p_k, rho_p)

# Calculation of the differential scattering cross-section "dscross"
# as a function of the scattering angle "theta" in the interval [0, \pi]
# in steps of 0.5 degree:

theta = scp.linspace(0.0, 180.0, 361)
ctheta = scp.cos(theta * scp.pi / 180.)
zi = 0. + 1.0j
pref = zi / k0
dscross = []
psi_s = []
for i in range(0, 361):
  psi = 0.0
  y = scs.lpn(ncut,ctheta[i])
  y1 = y[0]
  for n in range(0, ncut +1):
    d_n = (2 * n + 1)
    sum_n = - d_n * y1[n] * tm[n]
    psi = psi + sum_n
  psi = pref * psi
  psi_s = psi_s + [psi]
  dscross = dscross + [psi * scp.conj(psi)]
dscross = scp.real(dscross)

# Generation of the result file:

os.system("del dscross_centered_2l_sphere.txt")
os.system("type nul > dscross_centered_2l_sphere.txt")
fobj1 = open("dscross_centered_2l_sphere.txt", "w")
nl = "\n"
for i in range(0, 361):
  a1 = str(theta[i])
  a2 = " = "
  a3 = str(dscross[i]) + nl
  a = a1 + a2 + a3
  fobj1.write(a)
fobj1.close()

# Calculation of the total scattering cross-section "scat_tot" by use
# of the optical theorem:
```

```
print()
print()
print('Results: ')
w = scp.imag(psi_s[0])
scat_tot = 4 * scp.pi * w / k0
print()
print("total scattering cross-section: scat_tot = ", scat_tot)

# Plot of "dscross":

if plot_para == 'lg':
    plt.yscale('log')
    plt.plot(theta, dscross, color = 'black', linewidth=2.0)
else:
    plt.plot(theta, dscross, color = 'black', linewidth=2.0)
plt.xlabel("scattering angle (degree)", fontsize=16)
plt.ylabel("diff. scat. cross-sect.", fontsize=16)
plt.show()
```

```
#          Program "rotated_sphere_input.py"
#


print()
print('   --- Input scattering configuration for acoustic plane ...')
print('          ... wave scattering on a rotated sphere! ---')
print()

import os
os.system("del input_data_rotated_sphere.txt")
os.system("type nul > input_data_rotated_sphere.txt")
fobj = open("input_data_rotated_sphere.txt", "w")

para = str(input('soft (s), hard (h), or penetrable (p) sphere? '))
print()
if para == 'p':
    k_p_k = float(input('ratio of wave numbers k_p/k0: '))
    rho_p = float(input('density of the material inside the sphere rho_p: '))
else:
    k_p_k = 1.0
    rho_p = 1.0
a = float(input('radius a [mm]: '))
beta = float(input('size parameter beta: '))
print()
alpha = float(input('Eulerian angle alpha [degree]: '))
theta_p = float(input('Eulerian angle theta_p [degree]: '))
print()
ncut = int(input('truncation parameter ncut: '))
print()
plot_para = str(input('lin-lin (ll), or lin-log (lg) plot? '))

s0 = "para = " + para + "\n"
fobj.write(s0)
s1 = str(k_p_k)
s1 = "k_p_k = " + s1 + "\n"
fobj.write(s1)
s2 = str(rho_p)
s2 = "rho_p = " + s2 + "\n"
fobj.write(s2)
s3 = str(a)
s3 = "a = " + s3 + "\n"
fobj.write(s3)
```

```python
s4 = str(beta)
s4 = "beta = " + s4 + "\n"
fobj.write(s4)
s5 = str(alpha)
s5 = "alpha = " + s5 + "\n"
fobj.write(s5)
s6 = str(theta_p)
s6 = "theta_p = " + s6 + "\n"
fobj.write(s6)
s7 = str(ncut)
s7 = "ncut = " + s7 + "\n"
fobj.write(s7)
s8 = "plot_para = " + plot_para
fobj.write(s8)
fobj.close()
```

```python
#        Program "rotated_sphere.py"
#
# This program calculates the plane wave scattering behaviour of an
# acoustically soft, hard or penetrable sphere (homogeneous Dirichlet-,
# von Neumann- or mixed condition) arbitrarily rotated in the
# laboratory frame. Note that, if a penetrable sphere is considered,
# the density of the material "rho_0" outside the spheres is always given
# by 1.
#

import numpy as np
import scipy as scp
import scipy.special as scs
import matplotlib.pyplot as plt
import basics as bas
import os

# Read scattering configuration from input file:

fobj = open("input_data_rotated_sphere.txt", "r")

z = []
for line in fobj:
    line = line.strip()
    arr = line.split("= ")
    wert = str(arr[1])
    z = z + [wert]
fobj.close()

para = z[0]
k_p_k = float(z[1])
rho_p = float(z[2])
a = float(z[3])
beta = float(z[4])
alpha = float(z[5])
theta_p = float(z[6])
ncut = int(z[7])
plot_para = z[8]

k0 = beta / a
zi = 0. + 1.0j
```

```python
# Calculation of the T-matrix:

if para == 's':
   tm = bas.tm_s(ncut, beta)
elif para == 'h':
   tm = bas.tm_h(ncut, beta)
else:
   tm = bas.tm_p(ncut, beta, a, k0, k_p_k, rho_p)


# Calculation of the expansion coefficients in the laboratory frame:

c_n_l = []
for n in range(0, ncut + 1):
   d_n = np.sqrt(4. * np.pi * (2. * n + 1.)) * zi**n
   t_n = tm[n]
   c_l = []
   for l in range(-n, n + 1):
      c = 0.0
      for l1 in range(-n, n + 1):
         D_hin = bas.drm(n, l1, 0, 0., - theta_p, - alpha)
         D_rueck = bas.drm(n,l,l1, alpha, theta_p, 0.)
         c = c + d_n * t_n * D_hin * D_rueck
      c_l = c_l + [c]
   c_n_l = c_n_l + [c_l]


# Calculation of the differential scattering cross-section "dscross"
# as a function of the scattering angle "theta" in the interval [0, \pi]
# in steps of 1 degree:

theta = np.linspace(0.0, 180.0, 181)
r_theta = theta * scp.pi / 180.
ctheta = np.cos(r_theta)
pref = - zi / k0
dscross = []
psi_s = []
for i in range(0, 181):
   psi = 0.0
   for n in range(0, ncut + 1):
      c_n = c_n_l[n]
      for l in range(-n, n + 1):
         c_nl = c_n[l + n]
         Y_nl = scs.sph_harm(l,n,0.,r_theta[i])
         psi = psi + (-zi)**n * c_nl * Y_nl
```

```
    psi = pref * psi
    psi_s = psi_s + [psi]
    dscross = dscross + [psi * np.conj(psi)]
dscross = np.real(dscross)

# Generation of the result file:

os.system("del dscross_rotated_sphere.txt")
os.system("type nul > dscross_rotated_sphere.txt")
fobj1 = open("dscross_rotated_sphere.txt", "w")
nl = "\n"
for i in range(0, 181):
    a1 = str(theta[i])
    a2 = " = "
    a3 = str(dscross[i]) + nl
    a = a1 + a2 + a3
    fobj1.write(a)
fobj1.close()

# Calculation of the total scattering cross-section "scat_tot" by use
# of the optical theorem:

print()
print()
print('Results: ')
w = np.imag(psi_s[0])
scat_tot = 4 * scp.pi * w / k0
print()
print("total scattering cross-section: scat_tot = ", scat_tot)

# Plot of "dscross":

if plot_para == 'lg':
    plt.yscale('log')
    plt.plot(theta, dscross, color='black', linewidth=2.0)
else:
    plt.plot(theta, dscross, color='black', linewidth=2.0)
plt.xlabel("scattering angle [degree]", fontsize=16)
plt.ylabel("diff. scat. cross-sect.", fontsize=16)
plt.show()
```

```python
#         Program "z_shifted_sphere_input.py"
#

print()
print('      --- Input scattering configuration for acoustic plane ...')
print('          ... wave scattering on a z-shifted sphere! ---')
print()

import os
os.system("del input_data_z_shifted_sphere.txt")
os.system("type nul > input_data_z_shifted_sphere.txt")
fobj = open("input_data_z_shifted_sphere.txt", "w")

para = str(input('soft (s), hard (h), or penetrable (p) sphere? '))
print()
if para == 'p':
    k_p_k = float(input('ratio of wave numbers k_p/k0: '))
    rho_p = float(input('density of the material inside the sphere rho_p: '))
else:
    k_p_k = 1.0
    rho_p = 1.0
a = float(input('radius a [mm]: '))
beta = float(input('size parameter beta: '))
print()
b = float(input('shift along the z-axis b>=2a [mm]: '))
print()
ncut = int(input('truncation parameter ncut: '))
print()
plot_para = str(input('lin-lin (ll), or lin-log (lg) plot? '))

s0 = "para = " + para + "\n"
fobj.write(s0)
s1 = str(k_p_k)
s1 = "k_p_k = " + s1 + "\n"
fobj.write(s1)
s2 = str(rho_p)
s2 = "rho_p = " + s2 + "\n"
fobj.write(s2)
s3 = str(a)
s3 = "a = " + s3 + "\n"
fobj.write(s3)
s4 = str(beta)
s4 = "beta = " + s4 + "\n"
```

```
fobj.write(s4)
s5 = str(b)
s5 = "b = " + s5 + "\n"
fobj.write(s5)
s6 = str(ncut)
s6 = "ncut = " + s6 + "\n"
fobj.write(s6)
s7 = "plot_para = " + plot_para
fobj.write(s7)
fobj.close()
```

```
#          Program "z_shifted_sphere.py"
#
# This program calculates the plane wave scattering behaviour of a
# sound soft, hard or penetrable sphere (homogeneous Dirichlet-,
# von Neumann- or mixed condition) shifted along the z-axis of the
# laboratory frame. Note, that the density of the material "rho_0" outside
# the spheres is always given by 1. This program uses the expansion of
# the primary incident plane wave in the laboratory frame and its
# transformation into the shifted system afterwards.
#

import numpy as np
import scipy as scp
import scipy.special as scs
import matplotlib.pyplot as plt
import basics as bas
import os

# Read scattering configuration from input file:

fobj = open("input_data_z_shifted_sphere.txt", "r")

z = []
for line in fobj:
    line = line.strip()
    arr = line.split("= ")
    wert = str(arr[1])
    z = z + [wert]
fobj.close()

para = z[0]
k_p_k = float(z[1])
rho_p = float(z[2])
a = float(z[3])
beta = float(z[4])
b = float(z[5])
ncut = int(z[6])
plot_para = z[7]

k0 = beta / a
zi = 0. + 1.0j
k0b = k0 * b
```

```python
# Calculation of the T-matrix:

if para == 's':
    tm = bas.tm_s(ncut, beta)
elif para == 'h':
    tm = bas.tm_h(ncut, beta)
else:
    tm = bas.tm_p(ncut, beta, a, k0, k_p_k, rho_p)

# Calculation of the separation matrix:

y_os_n_nu = []
for i in range(0, ncut + 1):
    y_os_nu = []
    for j in range(0, ncut + 1):
        y_os_nu = y_os_nu + [bas.SM(0, i, j, k0b)]
    y_os_n_nu = y_os_n_nu + [y_os_nu]
y_o1_n_nu = np.real(y_os_n_nu)

# Calculation of the expansion coefficients in the laboratory frame:

def c_0_n_fun(nu_1, beta, kb, ncut):
    c = 0.0
    y_o_n = y_o1_n_nu[nu_1]
    for nu in range(0, ncut + 1):
        pref = (-1)**nu_1
        d_nu = np.sqrt(4. * np.pi * (2 * nu + 1)) * zi**nu
        S_rueck = y_o_n[nu]
        c = c + pref * S_rueck * tm[nu] * d_nu
    c = c * np.exp(zi * k0b)
    return c

# Calculation of the differential scattering cross-section "dscross"
# as a function of the scattering angle "theta" in the interval [0, \pi]
# in steps of 1 degree:

theta = np.linspace(0.0, 180.0, 181)
r_theta = theta * scp.pi / 180.
pf = - zi / k0
dscross = []
psi_s = []
print()
for i in range(0, 181):
```

```
      psi = 0.0
      for nu_1 in range(0, ncut + 1):
        c_ns = c_0_n_fun(nu_1, beta, k0b, ncut)
        Y_0_nu1 = scs.sph_harm(0,nu_1,0.,r_theta[i])
        sum = (-zi)**nu_1 * c_ns * Y_0_nu1
        psi = psi + sum
      psi = pf * psi
      psi_s = psi_s + [psi]
      dscross = dscross + [psi * np.conj(psi)]
  dscross = np.real(dscross)
  print()


# Generation of the result file:

  os.system("del dscross_z_shifted_sphere.txt")
  os.system("type nul > dscross_z_shifted_sphere.txt")
  fobj1 = open("dscross_z_shifted_sphere.txt", "w")
  nl = "\n"
  for i in range(0, 181):
      a1 = str(theta[i])
      a2 = " = "
      a3 = str(dscross[i]) + nl
      a = a1 + a2 + a3
      fobj1.write(a)
  fobj1.close()


# Calculation of the total scattering cross-section "scat_tot" by use
# of the optical theorem:

  print()
  print()
  print('Results: ')
  w = np.imag(psi_s[0])
  scat_tot = 4 * scp.pi * w / k0
  print()
  print("total scattering cross-section: scat_tot = ", scat_tot)


# Plot of "dscross":

  if plot_para == 'lg':
      plt.yscale('log')
      plt.plot(theta, dscross, color='black', linewidth=2.0)
  else:
```

```
    plt.plot(theta, dscross, color='black', linewidth=2.0)
plt.xlabel("scattering angle [degree]", fontsize=16)
plt.ylabel("diff. scat. cross-sect.", fontsize=16)
plt.show()
```

```
#        Program "shifted_sphere_input.py"
#

print()
print('   --- Input scattering configuration for acoustic plane ...')
print('         ... wave scattering on a shifted sphere! ---')
print()

import os
os.system("del input_data_shifted_sphere.txt")
os.system("type nul > input_data_shifted_sphere.txt")
fobj = open("input_data_shifted_sphere.txt", "w")

para = str(input('soft (s), hard (h), or penetrable (p) sphere? '))
print()
if para == 'p':
    k_p_k = float(input('ratio of wave numbers k_p/k0: '))
    rho_p = float(input('density of the material inside the sphere rho_p: '))
else:
    k_p_k = 1.0
    rho_p = 1.0
a = float(input('radius a [mm]: '))
beta = float(input('size parameter beta: '))
print()
alpha = float(input('Eulerian angle alpha [degree]: '))
theta_p = float(input('Eulerian angle theta_p [degree]: '))
b = float(input('shift along new z-axis  b>0 [mm]: '))
print()
ncut = int(input('truncation parameter ncut: '))
print()
plot_para = str(input('lin-lin (ll), or lin-log (lg) plot? '))

s0 = "para = " + para + "\n"
fobj.write(s0)
s1 = str(k_p_k)
s1 = "k_p_k = " + s1 + "\n"
fobj.write(s1)
s2 = str(rho_p)
s2 = "rho_p = " + s2 + "\n"
fobj.write(s2)
s3 = str(a)
s3 = "a = " + s3 + "\n"
fobj.write(s3)
```

```
s4 = str(beta)
s4 = "beta = " + s4 + "\n"
fobj.write(s4)
s5 = str(alpha)
s5 = "alpha = " + s5 + "\n"
fobj.write(s5)
s6 = str(theta_p)
s6 = "theta_p = " + s6 + "\n"
fobj.write(s6)
s7 = str(b)
s7 = "b = " + s7 + "\n"
fobj.write(s7)
fobj.close()
```

```
#        Program "shifted_sphere.py"
#
# Acoustic plane wave scattering on an arbitrarily shifted sphere.
# Calculation is not performed by use of a translation of the laboratory
# frame into the body frame but by using only a rotation and a shift
# along the new z-axis after the rotation! Note, that, if a penetrable
# sphere is considered, the density of the material "rho_0" outside the
# spheres is always given by 1.

#

print()
print()

import numpy as np
import scipy as scp
import scipy.special as scs
import matplotlib.pyplot as plt
import basics as bas
import os

# Read scattering configuration from input file:

fobj = open("input_data_shifted_sphere.txt", "r")

z = []
for line in fobj:
    line = line.strip()
    arr = line.split("= ")
    wert = str(arr[1])
    z = z + [wert]
fobj.close()

para = z[0]
k_p_k = float(z[1])
rho_p = float(z[2])
a = float(z[3])
beta = float(z[4])
alpha = float(z[5])
theta_p = float(z[6])
b = float(z[7])
ncut = int(z[8])
plot_para = z[9]
```

```
nangle = 91  # defines resolution in the scattering plane [0, \pi]
k0 = beta / a
k0b = k0 * b
zi = 0. + 1.0j


# Calculation of the separation matrix:

sm_l_i_j = []
for l in range(0, ncut + 1):
    sm_i_j = []
    for i in range(0, ncut + 1):
        sm_j = []
        for j in range(0, ncut + 1):
            sm_j = sm_j + [bas.SM(l, i, j, k0b)]
        sm_i_j = sm_i_j + [sm_j]
    sm_l_i_j = sm_l_i_j + [sm_i_j]
sm_l_i_j = np.real(sm_l_i_j)


# Calculation of the matrix of rotation required to perform
# the transformation into the body system:

dh_n_l = []
for l in range(-ncut, ncut + 1):
    dh_n = []
    for n in range(0, ncut + 1):
        dh_n = dh_n + [bas.drm(n, l, 0, 0., - theta_p, - alpha)]
    dh_n_l = dh_n_l + [dh_n]


# Calculation of the T-matrix:

if para == 's':
    tm = bas.tm_s(ncut, beta)
elif para == 'h':
    tm = bas.tm_h(ncut, beta)
else:
    tm = bas.tm_p(ncut, beta, a, k0, k_p_k, rho_p)


# Calculation of the differential scattering cross-section "dscross"
# in the laboratory frame as a function of the scattering angle "theta"
# in the interval [0, \pi] in steps depending on "nangle":

cmod_ls_nus = []
```

```python
for ls in range(-ncut, ncut + 1):
    cmod_nus = []
    for nus in range(np.abs(ls), ncut + 1):
        cmod = 0.0
        for l in range(-ncut, ncut + 1):
            y_l = sm_l_i_j[np.abs(l)]
            y_dh = dh_n_l[l + ncut]
            y_l_nus = y_l[nus]
            D_rueck = bas.drm(nus, ls, l, alpha, theta_p, 0.)
            csum = 0.0
            for en in range(np.abs(l), ncut + 1):
                pref = (-1)**en * zi**nus
                d_en = np.sqrt(4. * np.pi * (2 * en + 1)) * zi**en
                D_hin = y_dh[en]
                for nu in range(np.abs(l), ncut + 1):
                    S_rueck = y_l_nus[nu]
                    T_nu = tm[nu]
                    y_l_nu = y_l[nu]
                    S_hin = y_l_nu[en]
                    csum = csum + D_rueck * S_rueck * T_nu * S_hin * D_hin * \
                    d_en * pref
            cmod = cmod + csum
        cmod_nus = cmod_nus + [cmod]
    cmod_ls_nus = cmod_ls_nus + [cmod_nus]

theta = np.linspace(0.0, 180.0, nangle)
r_theta = theta * scp.pi / 180.
pf = - zi / k0
dscross = []
psi_s = []
print()
for i in range(0, nangle):
    psi = 0.0
    for ls in range(-ncut, ncut + 1):
        c_ls = cmod_ls_nus[ls + ncut]
        for nus in range(np.abs(ls), ncut + 1):
            nusz = nus - np.abs(ls)
            c_ls_nus = c_ls[nusz]
            Y_ls_nus = scs.sph_harm(ls,nus,0.,r_theta[i])
            psi = psi + c_ls_nus * Y_ls_nus
    psi = pf * psi
    psi_s = psi_s + [psi]
    dscross = dscross + [psi * np.conj(psi)]
```

```
dscross = np.real(dscross)
print()

# Generation of the result file:

os.system("del dscross_shifted_sphere.txt")
os.system("type nul > dscross_shifted_sphere.txt")
fobj1 = open("dscross_shifted_sphere.txt", "w")
nl = "\n"
for i in range(0, nangle):
    a1 = str(theta[i])
    a2 = " = "
    a3 = str(dscross[i]) + nl
    a = a1 + a2 + a3
    fobj1.write(a)
fobj1.close()

# Calculation of the total scattering cross-section "scat_tot" by use
# of the optical theorem:

print()
print()
print('Results: ')
w = np.imag(psi_s[0])
scat_tot = 4 * scp.pi * w / k0
print()
print("total scattering cross-section: scat_tot = ", scat_tot)

# Plot of "dscross":

if plot_para == 'lg':
    plt.yscale('log')
    plt.plot(theta, dscross, color='black', linewidth=2.0)
else:
    plt.plot(theta, dscross, color='black', linewidth=2.0)
plt.xlabel("scattering angle [degree]", fontsize=16)
plt.ylabel("diff. scat. cross-sect.", fontsize=16)
plt.show()
```

Appendix C
Full Python Programs of Chap. 3

© Springer Nature Switzerland AG 2020

T. Rother, *Sound Scattering on Spherical Objects*,

https://doi.org/10.1007/978-3-030-36448-9

```
#          Program "janus_spheres_ncut.py"
#
# This program aims at identifying the truncation parameter "ncut" for
# arbitrarily oriented h-p-, s-p-, or h-s-Janus spheres. This is accomplished
# by looking at the "ncut"-dependence of the total scattering cross-section
# of the respective Janus sphere in axisymmetric orientation. Note, that the
# density of the material "rho" outside the sphere is always given by 1.0.
#


print()
print()
print(" --- Total scattering cross-section of Janus spheres as a ---")
print("    ... function of ncut (axisymmetric orientation!). ---")
print()
print()


import scipy as scp
import scipy.special as scs
import basics as bas
import numpy as np
import os

# Input scattering configuration:

para = str(input('hard/soft (hs), soft/penetrable (sp), or hard/penetrable \
(hp) Janus sphere? '))
print()
if (para == 'sp' or para == 'hp'):
    k_p_k = float(input('ratio of wave numbers of Janus sphere k_p/k: '))
    rho_p = float(input('density of the material inside the Janus sphere \
    rho_p: '))
else:
    k_p_k = 1.0
    rho_p = 1.0
print()
a = float(input('radius a [mm]: '))
beta = float(input('size parameter beta: '))
theta_j = float(input('angle of coating theta_j: '))
print()
minnumber = int(input('lower limit of truncation parameter ncut: '))
maxnumber = int(input('upper limit of truncation parameter ncut: '))
print()
```

```
k0 = beta / a
kappa = rho_p / k_p_k
beta_p = k0 * k_p_k * a
zi = 0. + 1.0j
pref = - zi / k0


# Generation of the result file:

os.system("del scat_tot_ncut.txt")
os.system("type nul > scat_tot_ncut.txt")
fobj1 = open("scat_tot_ncut.txt", "w")
nl = "\n"    # new line

for ncut in range(minnumber, maxnumber + 1):
    print(ncut)
    d = []
    for n in range(0, ncut + 1):
        d = d + [np.sqrt(4. * np.pi * (2 * n + 1)) * zi**n]

# Calculation of the T-matrix:

    if para == 'hs':
        tm = bas.t_hs(0, beta, theta_j, ncut)
    elif para == 'sp':
        tm = bas.t_sp(0, beta, beta_p, theta_j, k_p_k, rho_p, ncut)
    else:
        tm = bas.t_hp(0, beta, beta_p, theta_j, k_p_k, rho_p, ncut)

# Calculation of the scattering coefficients:

    c = tm.dot(d)

# Calculation of the total scattering cross-section "scat_tot" by use
# of the optical theorem:

    psi_s = 0.0
    for n in range(0, ncut + 1):
        Y_0_n = scs.sph_harm(0,n,0.,0.)
        sum_n = c[0,n] * (-zi)**n * Y_0_n
        psi_s = psi_s + sum_n
    psi_s = pref * psi_s

    w = scp.imag(psi_s)
```

```
    scat_tot = 4 * scp.pi * w / k0
    scat_tot = str(scat_tot) + nl
    n_ncut = str(ncut)
    val = "ncut = " + n_ncut + " : " + " sigma_tot = " + scat_tot
    fobj1.write(val)
fobj1.close()
```

```python
#        Program "janus_axial_input.py"
print()
print(" --- Acoustic plane wave scattering on a Janus sphere! ---")
print("        ... (axisymmetric orientation) ---")
print()

import os
os.system("del input_data_janus_axial.txt")
os.system("type nul > input_data_janus_axial.txt")
fobj = open("input_data_janus_axial.txt", "w")

# Input scattering configuration:

para = str(input('hard-penetrable (hp), soft-penetrable (sp), or \
hard-soft (hs) Janus sphere? '))
print()
if para == 'hp' or para == 'sp':
    k_p_k = float(input('ratio of wave numbers k_p/k_0: '))
    rho_p = float(input('density of the material inside the sphere rho_p: '))
else:
    k_p_k = 1.0
    rho_p = 1.0
a = float(input('radius ... a [mm]: '))
beta = float(input('size parameter ... beta: '))
theta_j = float(input('splitting angle ... theta_j [degree]: '))
print()
ncut = int(input("truncation parameter ... ncut: "))
print()
plot_para = str(input('lin-lin (ll), or lin-log (lg) plot? '))

s0 = "para = " + para + "\n"
fobj.write(s0)

s1 = str(k_p_k)
s1 = "k_p_k = " + s1 + "\n"
fobj.write(s1)

s2 = str(rho_p)
s2 = "rho_p = " + s2 + "\n"
fobj.write(s2)

s3 = str(a)
s3 = "a = " + s3 + "\n"
fobj.write(s3)
```

```
s4 = str(beta)
s4 = "beta = " + s4 + "\n"
fobj.write(s4)

s5 = str(theta_j)
s5 = "theta_j = " + s5 + "\n"
fobj.write(s5)

s6 = str(ncut)
s6 = "ncut = " + s6 + "\n"
fobj.write(s6)

s7 = "plot_para = " + plot_para
fobj.write(s7)

fobj.close()
```

```python
#          Program "janus_axial.py"
#
# This program calculates the plane wave scattering behaviour of an
# axisymmetrically oriented Janus sphere that combines the boundary
# conditions of a sound hard, sound soft, and sound penetrable sphere
# in three different ways.
#

import scipy as scp
import scipy.special as scs
import matplotlib.pyplot as plt
import basics as bas
import numpy as np
import os

# Reading scattering configuration from input file:

fobj = open("input_data_janus_axial.txt", "r")

z = []
for line in fobj:
    line = line.strip()
    arr = line.split("= ")
    wert = str(arr[1])
    z = z + [wert]
fobj.close()

para = z[0]
k_p_k = float(z[1])
rho_p = float(z[2])
a = float(z[3])
beta = float(z[4])
theta_j = float(z[5])
ncut = int(z[6])
plot_para = z[7]

k = beta / a
kappa = rho_p / k_p_k
beta_p = k * k_p_k * a
theta = scp.linspace(0.0, 180.0, 181)
r_theta = theta * scp.pi / 180.
zi = 0. + 1.0j
pref = - zi / k
```

```
d = []
for n in range(0, ncut +1):
    d = d + [np.sqrt(4. * np.pi * (2 * n + 1)) * zi**n]


# Calculation of the T-matrix:

if para == 'hp':
    tm = bas.t_hp(0, beta, beta_p, theta_j, k_p_k, rho_p, ncut)
elif para == 'sp':
    tm = bas.t_sp(0, beta, beta_p, theta_j, k_p_k, rho_p, ncut)
else:
    tm = bas.t_hs(0, beta, theta_j, ncut)


# Calculation of the scattering coefficients:

c = tm.dot(d)


# Calculation of the differential scattering cross-section "dscross"
# as a function of the scattering angle "theta" in the interval [0, \pi]
# in steps of 1.0 degree:

dscross = []
psi_s = []
for i in range(0, 181):
    psi = 0.0
    for n in range(0, ncut +1):
        Y_0_n = scs.sph_harm(0,n,0.,r_theta[i])
        sum_n = c[0,n] * (-zi)**n * Y_0_n
        psi = psi + sum_n
    psi = pref * psi
    psi_s = psi_s + [psi]
    dscross = dscross + [psi * scp.conj(psi)]
dscross = scp.real(dscross)


# Generation of the result file:

os.system("del dscross_janus_axial.txt")
os.system("type nul > dscross_janus_axial.txt")
fobj1 = open("dscross_janus_axial.txt", "w")
nl = "\n"
for i in range(0, 181):
    a1 = str(theta[i])
```

```
    a2 = " = "
    a3 = str(dscross[i]) + nl
    a = a1 + a2 + a3
    fobj1.write(a)
fobj1.close()

# Calculation of the total scattering cross-section "scat_tot" by use
# of the optical theorem:

w = scp.imag(psi_s[0])
scat_tot = 4 * scp.pi * w / k
print()
print("total scattering cross-section: scat_tot = ", scat_tot)

# Plot of "dscross":

if plot_para == 'lg':
    plt.yscale('log')
    plt.plot(theta, dscross, color='black', linewidth=2.0)
else:
    plt.plot(theta, dscross, color='black', linewidth=2.0)
plt.xlabel("scattering angle [degree]", fontsize = 16)
plt.ylabel("diff. scat. cross-sect.", fontsize = 16)
plt.show()
```

```python
#        Program "janus_rotated_input.py"

print()
print(" --- Acoustic plane wave scattering on a Janus sphere! ---")
print("          ... (arbitrary rotation) ---")
print()

import os
os.system("del input_data_janus_rotated.txt")
os.system("type nul > input_data_janus_rotated.txt")
fobj = open("input_data_janus_rotated.txt", "w")

# Input scattering configuration:

para = str(input('hard-penetrable (hp), soft-penetrable (sp), or \
hard-soft (hs) Janus sphere? '))
print()
if para == 'hp' or para == 'sp':
   k_p_k = float(input('ratio of wave numbers k_p/k_0: '))
    rho_p = float(input('density of the material inside the sphere rho_p: '))
else:
   k_p_k = 1.0
   rho_p = 1.0
a = float(input('radius ... a [mm]: '))
beta = float(input('size parameter ... beta: '))
theta_j = float(input('splitting angle ... theta_j [degree]: '))
print()
alpha = float(input('Eulerian angle  alpha [degree]: '))
theta_p = float(input('Eulerian angle theta_p [Grad]: '))
print()
ncut = int(input("truncation parameter ...  ncut: "))
lcut = int(input("truncation parameter ...  lcut <= ncut: "))
print()
plot_para = str(input('lin-lin (ll), or lin-log (lg) plot? '))

s0 = "para = " + para + "\n"
fobj.write(s0)

s1 = str(k_p_k)
s1 = "k_p_k = " + s1 + "\n"
fobj.write(s1)

s2 = str(rho_p)
s2 = "rho_p = " + s2 + "\n"
```

```
fobj.write(s2)

s3 = str(a)
s3 = "a = " + s3 + "\n"
fobj.write(s3)

s4 = str(beta)
s4 = "beta = " + s4 + "\n"
fobj.write(s4)

s5 = str(theta_j)
s5 = "theta_j = " + s5 + "\n"
fobj.write(s5)

s6 = str(alpha)
s6 = "alpha = " + s6 + "\n"
fobj.write(s6)

s7 = str(theta_p)
s7 = "theta_p = " + s7 + "\n"
fobj.write(s7)

s8 = str(ncut)
s8 = "ncut = " + s8 + "\n"
fobj.write(s8)

s9 = str(lcut)
s9 = "lcut = " + s9 + "\n"
fobj.write(s9)

s10 = "plot_para = " + plot_para
fobj.write(s10)

fobj.close()
```

```python
#        Program "janus_rotated.py"
#
# This program calculates the plane wave scattering behaviour of an
# arbitrarily rotated Janus sphere that combines  the boundary
# conditions of a sound hard, sound soft, and sound penetrable sphere
# in three different ways.
#

import scipy as scp
import scipy.special as scs
import matplotlib.pyplot as plt
import basics as bas
import numpy as np
import os

# Reading scattering configuration from input file:

fobj = open("input_data_janus_rotated.txt", "r")

z = []
for line in fobj:
    line = line.strip()
    arr = line.split("= ")
    wert = str(arr[1])
    z = z + [wert]
fobj.close()

para = z[0]
k_p_k = float(z[1])
rho_p = float(z[2])
a = float(z[3])
beta = float(z[4])
theta_j = float(z[5])
alpha = float(z[6])
theta_p = float(z[7])
ncut = int(z[8])
lcut = int(z[9])
plot_para = z[10]

k = beta / a
kappa = rho_p / k_p_k
beta_p = k * k_p_k * a
zi = 0. + 1.0j
```

```
pref = - zi / k

# Calculation of the matrix of rotation for the transformation
# into the body frame:

dh_n_l = []
for l in range(-lcut, lcut + 1):
    dh_n = []
    for n in range(0, ncut + 1):
        dh_n = dh_n + [bas.drm(n, l, 0, 0., -theta_p, -alpha)]
    dh_n_l = dh_n_l + [dh_n]
dh_n_l = np.array(dh_n_l)


# Calculation of the expansion coefficients of the incident plane
# wave and the scattred field in the local system of the rotated
# Janus sphere:

d_l_n = []
for l in range(-lcut, lcut + 1):
    d_n = []
    for n in range(np.abs(l), ncut + 1):
        d_inc = np.sqrt(4. * np.pi * (2 * n + 1)) * zi**n
        d_inc = d_inc * dh_n_l[l + lcut, n]
        d_n = d_n + [d_inc]
    d_l_n = d_l_n + [d_n]


c_l_n = []
for l in range(-lcut, lcut + 1):
    print(' calculating T-matrix of l-mode l = ', l)
    d_l = d_l_n[l + lcut]
    d_l = np.array(d_l)
    if para == 'hp':
        tm_l = bas.t_hp(l, beta, beta_p, theta_j, k_p_k, rho_p, ncut)
    elif para == 'sp':
        tm_l = bas.t_sp(l, beta, beta_p, theta_j, k_p_k, rho_p, ncut)
    else:
        tm_l = bas.t_hs(l, beta, theta_j, ncut)
    cv = tm_l.dot(d_l)
    c_l_n = c_l_n + [cv]


# Calculation of the differential scattering cross-section "dscross"
# as a function of the scattering angle "theta" in the interval [0, \pi]
# in steps of 1.0 degree:
```

```
theta_l = 360.0
theta_l1 = 361
theta = np.linspace(0.0, theta_l, theta_l1)
r_theta = theta * scp.pi / 180.
dscross = []
psi_s = []
for i in range(0, theta_l1):
    print(' scattering angle theta = ', theta[i])
    psi = 0.0
    for l in range(-lcut, lcut + 1):
        cr_l = c_l_n[l + lcut]
        for n in range(np.abs(l), ncut + 1):
            nz = n - np.abs(l)
            cr = (-zi)**n * cr_l[0, nz]
            for ls in range(-n, n + 1):
                D_rueck = bas.drm(n, ls, l, alpha, theta_p, 0.)
                if i <= 180:
                    Y_ls_n = scs.sph_harm(ls,n,0.,r_theta[i])
                else:
                    Y_ls_n = scs.sph_harm(ls,n,np.pi,r_theta[360 - i])
                psi = psi + (cr * D_rueck * Y_ls_n)
    psi = pref * psi
    psi_s = psi_s + [psi]
    dscross = dscross + [psi * np.conj(psi)]
dscross = scp.real(dscross)

# Generation of the result file:

os.system("del dscross_janus_rotated.txt")
os.system("type nul > dscross_janus_rotated.txt")
fobj1 = open("dscross_janus_rotated.txt", "w")
nl = "\n"
for i in range(0, theta_l1):
    a1 = str(theta[i])
    a2 = " = "
    a3 = str(dscross[i]) + nl
    a = a1 + a2 + a3
    fobj1.write(a)
fobj1.close()

# Calculation of the total scattering cross-section "scat_tot" by use
# of the optical theorem:
```

```
w = scp.imag(psi_s[0])
scat_tot = 4 * scp.pi * w / k
print()
print("total scattering cross-section: scat_tot = ", scat_tot)

# Plot of "dscross":

if plot_para == 'lg':
    plt.yscale('log')
    plt.plot(theta, dscross, color='black', linewidth=2.0)
else:
    plt.plot(theta, dscross, color='black', linewidth=2.0)
plt.xlabel("scattering angle [degree]", fontsize=16)
plt.ylabel("diff. scat. cross-sect.", fontsize=16)
plt.show()
```

```python
#        Program "h_s_axial_surface.py"
#
# This program provide a test of the Dirichlet and Neumann condition of the
# surface field of an axisymmetrically oriented h-s-Janus sphere that combines
# the boundary conditions of a sound hard and soft sphere (homogeneous Neumann
# and Dirichlet boundary condition).
#

import scipy as scp
import scipy.special as scs
import matplotlib.pyplot as plt
import basics as bas
import numpy as np
import os

# Reading scattering configuration from input file:

fobj = open("input_data_h_s_axial.txt", "r")

z = []
for line in fobj:
    line = line.strip()
    arr = line.split("= ")
    wert = str(arr[1])
    z = z + [wert]
fobj.close()

a = float(z[0])
beta = float(z[1])
theta_j = float(z[2])
ncut = int(z[3])
plot_para = z[4]

k = beta / a
theta = scp.linspace(0.0, 180.0, 181)
r_theta = theta * scp.pi / 180.
zi = 0. + 1.0j
pref = - zi / k

d = []
for n in range(0, ncut +1):
    d = d + [np.sqrt(4. * np.pi * (2 * n + 1)) * zi**n]
```

```python
# Calculation of the T-matrix:

tm = bas.t_hs(0, beta, theta_j, ncut)

# Calculation of the scattering coefficients:

c = tm.dot(d)

# Calculation of the surface field (real and imaginary part) as a
# function of the scattering angle "theta" in the interval [0, \pi]
# in steps of 1.0 degree:

psi_t_d = []
psi_t_n = []
for i in range(0, 181):
    psi_scat_d = 0.0
    psi_inc_d = 0.0
    psi_scat_n = 0.0
    psi_inc_n = 0.0
    for n in range(0, ncut + 1):
        Y_0_n = scs.sph_harm(0,n,0.,r_theta[i])
        bes_d = scs.spherical_jn(n,beta)
        neu_d = scs.spherical_yn(n,beta)
        han_d = bes_d + zi * neu_d
        sum_n_scat_d = c[0,n] * han_d * Y_0_n
        sum_n_inc_d = d[n] * bes_d * Y_0_n
        psi_scat_d = psi_scat_d + sum_n_scat_d
        psi_inc_d = psi_inc_d + sum_n_inc_d
        bes_n = scs.spherical_jn(n,beta,derivative=True)
        neu_n = scs.spherical_yn(n,beta,derivative=True)
        han_n = bes_n + zi * neu_n
        sum_n_scat_n = c[0,n] * han_n * Y_0_n
        sum_n_inc_n = d[n] * bes_n * Y_0_n
        psi_scat_n = psi_scat_n + sum_n_scat_n
        psi_inc_n = psi_inc_n + sum_n_inc_n
    psi_t_n = psi_t_n + [psi_scat_n + psi_inc_n]
    psi_t_r_n = np.real(psi_t_n)
    psi_t_i_n = np.imag(psi_t_n)
    psi_t_d = psi_t_d + [psi_scat_d + psi_inc_d]
    psi_t_r_d = np.real(psi_t_d)
    psi_t_i_d = np.imag(psi_t_d)
```

```
# Generation of the result files:

os.system("del psi_tot_dirichlet_h_s_axial.txt")
os.system("type nul > psi_tot_dirichlet_h_s_axial.txt")
fobj1 = open("psi_tot_dirichlet_h_s_axial.txt", "w")
nl = "\n"
for i in range(0, 181):
    a1 = str(theta[i])
    a2 = " = "
    a3 = str(psi_t_d[i]) + nl
    a = a1 + a2 + a3
    fobj1.write(a)
fobj1.close()

os.system("del psi_tot_neumann_h_s_axial.txt")
os.system("type nul > psi_tot_neumann_h_s_axial.txt")
fobj2 = open("psi_tot_neumann_h_s_axial.txt", "w")
nl = "\n"
for i in range(0, 181):
    a1 = str(theta[i])
    a2 = " = "
    a3 = str(psi_t_n[i]) + nl
    a = a1 + a2 + a3
    fobj2.write(a)
fobj2.close()

# Plot of the surface field:

plt.figure(1)
plt.subplot(121)

plt.plot(theta, psi_t_r_d, color='black', linewidth=2.0, label = 'Dirichlet')
plt.xlabel("theta [degree]", fontsize = 16)
plt.ylabel("ext. tot. surface field (real)", fontsize = 16)
#plt.legend()
plt.legend(loc = 'upper right')

plt.subplot(122)

plt.plot(theta, psi_t_i_d, color='black', linewidth=2.0, label = 'Dirichlet')
plt.xlabel("theta [degree]", fontsize = 16)
plt.ylabel("ext. tot. surface field (imag)", fontsize = 16)
#plt.legend()
```

```
plt.legend(loc = 'upper right')

print(' Close graphic window to continue with test of Neumann condition')
plt.show()

plt.figure(1)
plt.subplot(121)

plt.plot(theta, psi_t_r_n, color='black', linewidth=2.0, label = 'Neumann')
plt.xlabel("theta [degree]", fontsize = 16)
plt.ylabel("ext. tot. surface field (real)", fontsize = 16)
#plt.legend()
plt.legend(loc = 'upper right')

plt.subplot(122)

plt.plot(theta, psi_t_i_n, color='black', linewidth=2.0, label = 'Neumann')
plt.xlabel("theta [degree]", fontsize = 16)
plt.ylabel("ext. tot. surface field (imag)", fontsize = 16)
#plt.legend()
plt.legend(loc = 'upper right')
plt.show()
```

Appendix D
Full Python Programs of Chap. 4

© Springer Nature Switzerland AG 2020
T. Rother, *Sound Scattering on Spherical Objects*,
https://doi.org/10.1007/978-3-030-36448-9

```python
#          Program "bisphere_approxy_input.py"
#

print()
print('   --- Input scattering configuration for acoustic plane ...')
print('    ... wave scattering on combinations of sound soft ...')
print("        ... hard, or penetrable spheres arbitrarily ...")
print("   ... oriented in the laboratory frame (no interaction is ...")
print("                        ... considered!) ---")
print()

import os
os.system("del input_data_bisphere_approxy.txt")
os.system("type nul > input_data_bisphere_approxy.txt")
fobj = open("input_data_bisphere_approxy.txt", "w")

para0 = str(input('centered soft (s), hard (h) or penetrable (p) sphere? '))
para1 = str(input('shifted soft (s), hard (h) or penetrable (p) sphere? '))

if (para0 == 'p' and (para1 == 's' or para1 == 'h')):
    k_p_k_0 = float(input('ratio of wave numbers of centered sphere k_p0/k: '))
    rho_p_0 = float(input('density of the material inside the centered sphere \
    rho_p0: '))
elif (para1 == 'p' and (para0 == 's' or para0 == 'h')):
    k_p_k_1 = float(input('ratio of wave numbers of shifted sphere k_p1/k: '))
    rho_p_1 = float(input('density of the material inside the shifted sphere \
    rho_p1: '))
elif (para0 == 'p' and para1 == 'p'):
    k_p_k_0 = float(input('ratio of wave numbers of centered sphere k_p0/k: '))
    rho_p_0 = float(input('density of the material inside the centered sphere \
    rho_p0: '))
    k_p_k_1 = float(input('ratio of wave numbers of shifted sphere k_p1/k: '))
    rho_p_1 = float(input('density of the material inside the shifted sphere \
    rho_p1: '))
else:
    k_p_k_0 = 1.0
    rho_p_0 = 1.0
    k_p_k_1 = 1.0
    rho_p_1 = 1.0
print()
a0 = float(input('centered sphere with radius a0 [mm]: '))
a1 = float(input('shifted sphere with radius a1 [mm]: '))
beta_0 = float(input('size parameter of centered sphere beta_0: '))
```

```python
print()
b = float(input('distance between the centers of the spheres b [mm]: '))
alpha = float(input('Eulerian angle of orientation alpha [deg.]: '))
theta_p = float(input('Eulerian angle of orientation theta_p [deg]: '))
print()
ppm = int(input('dif. scat. cross. between [0,pi] (0) or [0,2*pi] (1)? '))
plot_para = str(input('lin-lin (ll), or lin-log (lg) plot? '))

s0 = "para0 = " + para0 + "\n"
fobj.write(s0)

s1 = "para1 = " + para1 + "\n"
fobj.write(s1)

s2 = str(k_p_k_0)
s2 = "k_p_k_0 = " + s2 + "\n"
fobj.write(s2)

s3 = str(rho_p_0)
s3 = "rho_p_0 = " + s3 + "\n"
fobj.write(s3)

s4 = str(k_p_k_1)
s4 = "k_p_k_1 = " + s4 + "\n"
fobj.write(s4)

s5 = str(rho_p_1)
s5 = "rho_p_1 = " + s5 + "\n"
fobj.write(s5)

s6 = str(a0)
s6 = "a0 = " + s6 + "\n"
fobj.write(s6)

s7 = str(a1)
s7 = "a1 = " + s7 + "\n"
fobj.write(s7)

s8 = str(beta_0)
s8 = "beta_0 = " + s8 + "\n"
fobj.write(s8)

s9 = str(b)
```

```
s9 = "b = " + s9 + "\n"
fobj.write(s9)

s10 = str(alpha)
s10 = "alpha = " + s10 + "\n"
fobj.write(s10)

s11 = str(theta_p)
s11 = "theta_p = " + s11 + "\n"
fobj.write(s11)

s12 = str(ppm)
s12= "ppm = " + s12 + "\n"
fobj.write(s12)

s13 = "plot_para = " + plot_para
fobj.write(s13)

fobj.close()
```

```python
#        Program "bisphere_approxy.py"
#
# This program approximates the plane wave scattering behaviour of any
# combination of  sound soft, hard or penetrable spheres that
# are arbitrarily oriented in the laboratory frame! This approximation
# neglects any interaction between the spheres and consideres only the
# interference term. The configuration is fixed by the distance "b", and
# the Eulerian angles "alpha" and "\theta_p".
#

import numpy as np
import scipy as scp
import scipy.special as scs
import matplotlib.pyplot as plt
import basics as bas
import os

# Read scattering configuration from input file:

fobj = open("input_data_bisphere_approxy.txt", "r")

z = []
for line in fobj:
    line = line.strip()
    arr = line.split("= ")
    wert = str(arr[1])
    z = z + [wert]
fobj.close()

para0 = z[0]
para1 = z[1]
k_p_k_0 = float(z[2])
rho_p_0 = float(z[3])
k_p_k_1 = float(z[4])
rho_p_1 = float(z[5])
a0 = float(z[6])
a1 = float(z[7])
beta_0 = float(z[8])
b = float(z[9])
alpha = float(z[10])
theta_p = float(z[11])
ppm = int(z[12])
plot_para = z[13]
```

```python
# Fixing angular resolution and truncation parameter:

if ppm == 1:
    theta_l = 360.0
    theta_l1 = 361
    theta = np.linspace(0.0, theta_l, theta_l1)
    ctheta = np.cos(theta * scp.pi / 180.)
    ctheta[0] = 1.0
    ctheta[180] = -1.0
else:
    theta_l = 180.0
    theta_l1 = 181
    theta = np.linspace(0.0, theta_l, theta_l1)
    ctheta = np.cos(theta * scp.pi / 180.)
    ctheta[0] = 1.0


n0cut = int(beta_0 + 9.) # may be changed for larger size parameters!
k = beta_0 / a0
kb = k * b
beta_1 = k * a1
n1cut = int(beta_1 + 9.) # may be changed for larger size parameters!


# Calculation of the T-matrices:

if (para0 == 's' and para1 == 's'):
    tm_0 = bas.tm_s(n0cut, beta_0)
    tm_1 = bas.tm_s(n1cut, beta_1)
elif (para0 == 's' and para1 == 'h'):
    tm_0 = bas.tm_s(n0cut, beta_0)
    tm_1 = bas.tm_h(n1cut, beta_1)
elif (para0 == 's' and para1 == 'p'):
    tm_0 = bas.tm_s(n0cut, beta_0)
    tm_1 = bas.tm_p(n1cut, beta_1, a1, k, k_p_k_1, rho_p_1)
elif (para0 == 'h' and para1 == 's'):
    tm_0 = bas.tm_h(n0cut, beta_0)
    tm_1 = bas.tm_s(n1cut, beta_1)
elif (para0 == 'h' and para1 == 'h'):
    tm_0 = bas.tm_h(n0cut, beta_0)
    tm_1 = bas.tm_h(n1cut, beta_1)
elif (para0 == 'h' and para1 == 'p'):
    tm_0 = bas.tm_h(n0cut, beta_0)
    tm_1 = bas.tm_p(n1cut, beta_1, a1, k, k_p_k_1, rho_p_1)
elif (para0 == 'p' and para1 == 's'):
```

```python
    tm_0 = bas.tm_p(n0cut, beta_0, a0, k, k_p_k_0, rho_p_0)
    tm_1 = bas.tm_s(n1cut, beta_1)
elif (para0 == 'p' and para1 == 'h'):
    tm_0 = bas.tm_p(n0cut, beta_0, a0, k, k_p_k_0, rho_p_0)
    tm_1 = bas.tm_h(n1cut, beta_1)
else:
    tm_0 = bas.tm_p(n0cut, beta_0, a0, k, k_p_k_0, rho_p_0)
    tm_1 = bas.tm_p(n1cut, beta_1, a1, k, k_p_k_1, rho_p_1)


# Calculation of the differential scattering cross-section as a
# function of the scattering angle "theta" in the interval [0, \pi]
# or [0, 2 * \pi] in steps of 1 degree:

zi = 0. + 1.0j
pref = -zi / k
dscross = []
psi_s = []
for i in range(0, theta_l1):
    psi0_1 = 0.0
    psi1_1 = 0.0
    y0 = scs.lpn(n0cut,ctheta[i])
    y1 = scs.lpn(n1cut,ctheta[i])
    y0_1 = y0[0]
    y1_1 = y1[0]
    for n0 in range(0, n0cut + 1):
        d_n0 = (2 * n0 + 1)
        sum_n0 = d_n0 * y0_1[n0] * tm_0[n0]
        psi0_1 = psi0_1 + sum_n0
    for n1 in range(0, n1cut + 1):
        d_n1 = (2 * n1 + 1)
        sum_n1 = d_n1 * y1_1[n1] * tm_1[n1]
        psi1_1 = psi1_1 + sum_n1
    psi0_1 = pref * psi0_1
    psi1_1 = pref * psi1_1
    c_k = kb * (np.cos(theta[i] * scp.pi / 180.) * np.cos(theta_p * \
            scp.pi / 180.) + np.sin(theta[i] * scp.pi / 180.) * \
    np.sin(theta_p * scp.pi / 180.) * np.cos(alpha *scp.pi / 180.))
    c_p = kb * np.cos(theta_p * scp.pi / 180.)
    psi1_1 = psi1_1 * np.exp(zi * (c_p - c_k))
    psi = psi0_1 + psi1_1
    psi_s = psi_s + [psi]
    dscross = dscross + [psi * np.conj(psi)]
dscross = np.real(dscross)
```

```
# Generation of the result file:

os.system("del dscross_bisphere_approxy.txt")
os.system("type nul > dscross_bisphere_approxy.txt")
fobj1 = open("dscross_bisphere_approxy.txt", "w")
nl = "\n"
for i in range(0, theta_l1):
    a1 = str(theta[i])
    a2 = " = "
    a3 = str(dscross[i]) + nl
    a = a1 + a2 + a3
    fobj1.write(a)
fobj1.close()

# Calculation of the total scattering cross-section "scat_tot" by use
# of the optical theorem:

print()
print()
print('Results: ')
w = np.imag(psi_s[0])
scat_tot = 4 * scp.pi * w / k
print()
print("total scattering cross-section: scat_tot = ", scat_tot)
print()

# Plot of "dscross":

if plot_para == 'lg':
    plt.yscale('log')
    plt.plot(theta, dscross, color = 'black', linewidth=2.0)
else:
    plt.plot(theta, dscross, color = 'black', linewidth=2.0)
plt.xlabel("scattering angle [degree]",fontsize=16)
plt.ylabel("diff. scat. cross-sect.",fontsize=16)
plt.show()
```

```python
#         Program "bisphere_z_oriented_input.py"
#


print()
print(" --- Input scattering configuration for acoustic plane wave  ...")
print("  ... scattering on a z-oriented bisphere. Note that the sphere  ...")
print("    ... with radius r=a0 is centered in the laboratoy frame. ---")
print()

import os
os.system("del input_data_bisphere_z_oriented.txt")
os.system("type nul > input_data_bisphere_z_oriented.txt")
fobj = open("input_data_bisphere_z_oriented.txt", "w")



LS = int(input(' 0. iteration (0), 1. iteratio (1), 2. iteration (2) or \
exact solution (3)? '))
print()
para0 = str(input('centered soft (s), hard (h) or penetrable (p) sphere? '))
para1 = str(input('shifted soft (s), hard (h) or penetrable (p) sphere? '))

if (para0 == 'p' and (para1 == 's' or para1 == 'h')):
    k_p_k_0 = float(input('ratio of wave numbers of centered sphere k_p0/k_0: '))
    rho_p_0 = float(input('density of the material inside the centered sphere \
    rho_p0: '))
elif (para1 == 'p' and (para0 == 's' or para0 == 'h')):
    k_p_k_1 = float(input('ratio of wave numbers of shifted sphere k_p1/k_0: '))
    rho_p_1 = float(input('density of the material inside the shifted sphere \
    rho_p1: '))
elif (para0 == 'p' and para1 == 'p'):
    k_p_k_0 = float(input('ratio of wave numbers of centered sphere k_p0/k_0: '))
    rho_p_0 = float(input('density of the material inside the centered sphere \
    rho_p0: '))
    k_p_k_1 = float(input('ratio of wave numbers of shifted sphere k_p1/k_0: '))
    rho_p_1 = float(input('density of the material inside the shifted sphere \
    rho_p1: '))
else:
    k_p_k_0 = 1.0
    rho_p_0 = 1.0
    k_p_k_1 = 1.0
    rho_p_1 = 1.0
print()
a0 = float(input('centered sphere with radius a0 [mm]: '))
```

```
a1 = float(input('shifted sphere with radius a1 [mm]: '))
beta_0 = float(input('size parameter of centered sphere beta_0: '))
print()
b = float(input('distance between the centers of the spheres b [mm]: '))
print()
ncut = int(input('truncation parameter ... ncut: '))
print()
plot_para = str(input('lin-lin (ll), or lin-log (lg) plot? '))
pl_pm1 = str(input("additional plot of Mie results (y/n)? "))


s00 = str(LS)
s00 = "LS = " + s00 + "\n"
fobj.write(s00)


s0 = "para0 = " + para0 + "\n"
fobj.write(s0)


s1 = "para1 = " + para1 + "\n"
fobj.write(s1)


s2 = str(k_p_k_0)
s2 = "k_p_k_0 = " + s2 + "\n"
fobj.write(s2)


s3 = str(rho_p_0)
s3 = "rho_p_0 = " + s3 + "\n"
fobj.write(s3)


s4 = str(k_p_k_1)
s4 = "k_p_k_1 = " + s4 + "\n"
fobj.write(s4)


s5 = str(rho_p_1)
s5 = "rho_p_1 = " + s5 + "\n"
fobj.write(s5)


s6 = str(a0)
s6 = "a0 = " + s6 + "\n"
fobj.write(s6)


s7 = str(a1)
s7 = "a1 = " + s7 + "\n"
fobj.write(s7)
```

```python
s8 = str(beta_0)
s8 = "beta_0 = " + s8 + "\n"
fobj.write(s8)

s9 = str(b)
s9 = "b = " + s9 + "\n"
fobj.write(s9)

s10 = str(ncut)
s10 = "ncut = " + s10 + "\n"
fobj.write(s10)

s11 = "plot_para = " + plot_para + "\n"
fobj.write(s11)

s12 = "pl_pm1 = " + pl_pm1
fobj.write(s12)

fobj.close()
```

```python
#         Program "bisphere_z_oriented.py"
#
# Acoustic plane wave scattering on a z-oriented bisphere (Iteration and
# rigorous solution). Note that the  sphere with radius "r=a0" is centered
# in the laboratory frame.
#
print()

import numpy as np
import scipy as scp
import scipy.special as scs
import scipy.linalg as scl
import matplotlib.pyplot as plt
import basics as bas
import os

# Reading scattering configuration from input file:

fobj = open("input_data_bisphere_z_oriented.txt", "r")

z = []
for line in fobj:
    line = line.strip()
    arr = line.split("= ")
    wert = str(arr[1])
    z = z + [wert]
fobj.close()

LS = int(z[0])
para0 = z[1]
para1 = z[2]
k_p_k_0 = float(z[3])
rho_p_0 = float(z[4])
k_p_k_1 = float(z[5])
rho_p_1 = float(z[6])
a0 = float(z[7])
a1 = float(z[8])
beta_0 = float(z[9])
b = float(z[10])
ncut = int(z[11])
plot_para = z[12]
pl_pm1 = z[13]
```

```
n1cut = ncut
theta_p = 0.0
zi = 0. + 1.0j
k = beta_0 / a0
kb = k * b
beta_1 = k * a1

# Calculation of the T-matrices:

if (para0 == 's' and para1 == 's'):
    tm_0 = bas.tm_s(ncut, beta_0)
    tm_1 = bas.tm_s(n1cut, beta_1)
elif (para0 == 's' and para1 == 'h'):
    tm_0 = bas.tm_s(ncut, beta_0)
    tm_1 = bas.tm_h(n1cut, beta_1)
elif (para0 == 's' and para1 == 'p'):
    tm_0 = bas.tm_s(ncut, beta_0)
    tm_1 = bas.tm_p(n1cut, beta_1, a1, k, k_p_k_1, rho_p_1)
elif (para0 == 'h' and para1 == 's'):
    tm_0 = bas.tm_h(ncut, beta_0)
    tm_1 = bas.tm_s(n1cut, beta_1)
elif (para0 == 'h' and para1 == 'h'):
    tm_0 = bas.tm_h(ncut, beta_0)
    tm_1 = bas.tm_h(n1cut, beta_1)
elif (para0 == 'h' and para1 == 'p'):
    tm_0 = bas.tm_h(ncut, beta_0)
    tm_1 = bas.tm_p(n1cut, beta_1, a1, k, k_p_k_1, rho_p_1)
elif (para0 == 'p' and para1 == 's'):
    tm_0 = bas.tm_p(ncut, beta_0, a0, k, k_p_k_0, rho_p_0)
    tm_1 = bas.tm_s(n1cut, beta_1)
elif (para0 == 'p' and para1 == 'h'):
    tm_0 = bas.tm_p(ncut, beta_0, a0, k, k_p_k_0, rho_p_0)
    tm_1 = bas.tm_h(n1cut, beta_1)
else:
    tm_0 = bas.tm_p(ncut, beta_0, a0, k, k_p_k_0, rho_p_0)
    tm_1 = bas.tm_p(n1cut, beta_1, a1, k, k_p_k_1, rho_p_1)

# Calculation of the separation matrix:

y_m0_m1 = []
for i in range(0, ncut + 1):
    y_os = []
    for j in range(0, ncut + 1):
```

```
      y_os = y_os + [bas.SM(0, i, j, kb)]
    y_m0_m1 = y_m0_m1 + [y_os]
```

```
# Mie calculation for single spheres:
```

```
c_m0_0 = []
for m0 in range(0, ncut + 1):
    d_m0 = np.sqrt(4. * np.pi * (2 * m0 + 1)) * zi**m0
    c = tm_0[m0] * d_m0
    c_m0_0 = c_m0_0 + [c]
```

```
q_m1_0 = []
for m1 in range(0, n1cut + 1):
    d_m1 = np.sqrt(4. * np.pi * (2 * m1 + 1)) * zi**m1
    q = tm_1[m1] * d_m1
    q_m1_0 = q_m1_0 + [q]
```

```
# Calculation of the matrices of the T-matrix equation:
```

```
c11 = []
for ic11 in range(0, ncut + 1):
    c1 = []
    for jc11 in range(0, ncut + 1):
        c = (1.0 if jc11 == ic11 else 0.0)
        c1 = c1 + [c]
    c11 = c11 + [c1]
```

```
q22 = []
for iq22 in range(0, n1cut + 1):
    q2 = []
    for jq22 in range(0, n1cut + 1):
        q = (1.0 if jq22 == iq22 else 0.0)
        q2 = q2 + [q]
    q22 = q22 + [q2]
```

```
q12 = []
c0 = []
for iq12 in range(0, ncut + 1):
    q1 = []
    S_m0 = y_m0_m1[iq12]
    q_to = - tm_0[iq12] * (-1)**iq12
    d_ic = np.sqrt(4. * np.pi * (2 * iq12 + 1)) * zi**iq12
    c0 = c0 + [tm_0[iq12] * d_ic]
```

```python
    for jq12 in range(0, n1cut + 1):
        q = q_to * S_m0[jq12]
        q1 = q1 + [q]
    q12 = q12 + [q1]


c21 = []
q0 = []
for ic21 in range(0, n1cut + 1):
    c2 = []
    S_m0 = y_m0_m1[ic21]
    c_to1 = - tm_1[ic21]
    d_iq = np.sqrt(4. * np.pi * (2 * ic21 + 1)) * zi**ic21
    q0 = q0 + [np.exp(zi * kb) * tm_1[ic21] * d_iq]
    for jc21 in range(0, ncut + 1):
        c = c_to1 * S_m0[jc21] * (-1)**jc21
        c2 = c2 + [c]
    c21 = c21 + [c2]


c11 = np.array(c11)
q22 = np.array(q22)
c21 = np.array(c21)
q12 = np.array(q12)
c0 = np.array(c0)
q0 = np.array(q0)


cq = np.matmul(c21, q12)
qc = np.matmul(q12, c21)


# Solving the T-matrix equation (iteration and rigorous):

if LS == 0:
    mc = c11
    mq = q22
    ci = c0
    qi = q0
    c_m0_00 = scl.solve(mc,ci)
    q_m1_00 = scl.solve(mq,qi)
elif LS == 1:
    mc = c11
    mq = q22
    ci = c0 - q12.dot(q0)
    qi = q0 - c21.dot(c0)
    c_m0_00 = scl.solve(mc,ci)
```

```
    q_m1_00 = scl.solve(mq,qi)
  elif LS == 2:
    mc = c11
    mq = q22
    ci1 = c0 - q12.dot(q0)
    qi1 = q0 - c21.dot(c0)
    ci_zw = np.matmul(c21, ci1)
    ci = ci1 + np.matmul(q12, ci_zw)
    qi_zw = np.matmul(q12, qi1)
    qi = qi1 + np.matmul(c21, qi_zw)
    c_m0_00 = scl.solve(mc,ci)
    q_m1_00 = scl.solve(mq,qi)
  elif LS == 3:
    mc = c11 - qc
    mq = q22 - cq
    ci = c0 - q12.dot(q0)
    qi = q0 - c21.dot(c0)
    c_m0_00 = scl.solve(mc,ci)
    q_m1_00 = scl.solve(mq,qi)
  c_m0 = c_m0_00
  q_m1 = q_m1_00


# Calculation of the differential scattering cross-section "dscross_g"
# as a function of the scattering angle "theta" in the interval [0, \pi]
# in steps of 1 degree.

theta_l = 180.0
theta_l1 = 181
theta = np.linspace(0.0, theta_l, theta_l1)
r_theta = theta * scp.pi / 180.
ctheta = np.cos(r_theta)
zi = 0. + 1.0j
pref = - zi / k
psi_s_g = []
dscross_g = []
dscross_0 = []
dscross_1 = []
for i in range(0, 181):
    psi_0 = 0.0
    psi_0_1 = 0.0
    psi_1 = 0.0
    psi_0_0 = 0.0
```

```python
    for m0 in range(0, ncut + 1):
        Y_0m0 = scs.sph_harm(0,m0,0.,r_theta[i])
        psi_0 = psi_0 + c_m0[m0] * (-zi)**m0 * Y_0m0
        psi_0_0 = psi_0_0 + c_m0_0[m0] * (-zi)**m0 * Y_0m0
    psi_0 = pref * psi_0
    psi_0_0 = pref * psi_0_0
    dscross_0 = dscross_0 + [psi_0_0 * np.conj(psi_0_0)]

    for m1 in range(0, n1cut + 1):
        Y_0m1 = scs.sph_harm(0,m1,0.,r_theta[i])
        psi_0_1 = psi_0_1 + q_m1_0[m1] * (-zi)**m1 * Y_0m1
    psi_0_1 = pref * psi_0_1
    dscross_1 = dscross_1 + [psi_0_1 * np.conj(psi_0_1)]

    for m1s in range(0, n1cut + 1):
        Y_0m1s = scs.sph_harm(0,m1s,0.,r_theta[i])
        sum_m1s = q_m1[m1s] * (-zi)**m1s * Y_0m1s
        psi_1 = psi_1 + sum_m1s
    c_k = kb * np.cos(np.abs(theta_p - theta[i]) * scp.pi / 180.)
    psi_1 = pref * psi_1 * np.exp(- zi * c_k)
    psi_s = psi_0 + psi_1
    psi_s_g = psi_s_g + [psi_s]
    dscross_g = dscross_g + [psi_s * np.conj(psi_s)]

dscross_g = np.real(dscross_g)
dscross_0 = np.real(dscross_0)
dscross_1 = np.real(dscross_1)
dscross_zw = 2. * (dscross_0 + dscross_1)
dscross_zw1 = dscross_0 + dscross_1  # für a1 << a0 !

# Generation of the result file:

os.system("del dscross_bisphere_z_oriented.txt")
os.system("type nul > dscross_bisphere_z_oriented.txt")
fobj1 = open("dscross_bisphere_z_oriented.txt", "w")
nl = "\n"
for i in range(0, theta_l1):
    a1 = str(theta[i])
    a2 = " = "
    a3 = str(dscross_g[i]) + nl
    a = a1 + a2 + a3
    fobj1.write(a)
fobj1.close()
```

```python
# Calculation of the total scattering cross-section "scat_tot" by use
# of the optical theorem:

print()
print()
print('Results: ')
w = np.imag(psi_s_g[0])
scat_tot = 4 * scp.pi * w / k
print()
print("total scattering cross-section: scat_tot = ", scat_tot)
print()


# Plot of the results:

if plot_para == 'lg':
    plt.yscale('log')
    if pl_pm1 == 'y':
        plt.plot(theta, dscross_zw, '-.', color = 'black', linewidth=2.0, \
                label = "2 * (Mie_a0 + Mie_a1)")
        plt.plot(theta, dscross_g, color='black', linewidth=2.0, \
                label = "bisphere")
    else:
        plt.plot(theta, dscross_g, linewidth=2.0, label = "bisphere")
else:
    if pl_pm1 == 'y':
        plt.plot(theta, dscross_zw, '-.', color = 'black', linewidth=2.0, \
                label = "2 * (Mie_a0 + Mie_a1)")
        plt.plot(theta, dscross_g, color='black', linewidth=2.0, \
                label = "bisphere")
    else:
        plt.plot(theta, dscross_g, color = 'black', linewidth=2.0, \
                label = "bisphere")
plt.xlabel("scattering angle [degree]", fontsize=16)
plt.ylabel("diff. scat. cross-sect.", fontsize=16)
plt.legend(loc = 'upper center')
# plt.legend()
plt.show()
```

```
#         Program "sumtest_1.py"
#
# This program provides a numerical test of (4.31) and (4.32) in Chpt. 4.
#

print()
print("              --- Sum test 1! ---")
print()

import numpy as np
import scipy as scp
import scipy.special as scs
import matplotlib.pyplot as plt
import basics as bas

# Input:

kb = float(input(' real; shift ... k_0*b : '))
m1 = int(input(' integer; mode ... m0 : '))
l1 = int(input(' integer; mode ... l1 : '))
ncut = int(input(' integer; truncation parameter ... ncut : '))

# Calculation of the analytical function and the approximation:

zi = 0. + 1.0j
y_m0_m1 = []
for i in range(0, ncut + 1):
    y_os = []
    for j in range(0, ncut + 1):
        y_os = y_os + [bas.SM(l1, i, j, kb)]
    y_m0_m1 = y_m0_m1 + [y_os]
y_m0_m1 = np.real(y_m0_m1)
S_m1 = y_m0_m1[m1]

theta = np.linspace(0.0, 180.0, 181)
r_theta = theta * scp.pi / 180.

sum_phase = []
ana_phase = []
for i in range(0, 181):
    c_k = kb * np.cos(r_theta[i])
    Y_l1_m1 = scs.sph_harm(l1,m1,0.,r_theta[i])
    sum = 0.0
```

```
        abs_l1 = np.abs(l1)
        for m1s in range(abs_l1, ncut + 1):
            Y_l1_m1s = scs.sph_harm(l1,m1s,0.,r_theta[i])
            sum = sum + zi**(m1s + m1) * S_m1[m1s] * Y_l1_m1s
        sum_phase = sum_phase + [sum]
        ana = np.exp(- zi * c_k) * Y_l1_m1
        ana_phase = ana_phase + [ana]


f1_r = np.real(ana_phase)
f2_r = np.real(sum_phase)


f1_i = np.imag(ana_phase)
f2_i = np.imag(sum_phase)


plt.figure(1)
plt.subplot(121)
#plt.yscale('log')
plt.plot(theta, f1_r, color = 'black', linewidth=2.0, label = "analytical")
plt.plot(theta[0], f2_r[0], 'ro', linewidth=2.0, color = 'black', label = "sum")
for i in range(1, 18):
    plt.plot(theta[10 * i + 1], f2_r[10 * i + 1], 'ro', \
         linewidth=2.0, color = 'black')
plt.plot(theta[180], f2_r[180], 'ro', linewidth=2.0, color = 'black')
#plt.plot(theta, f2_r, color = 'black', linestyle = 'dashed', linewidth=2.0, \
#      label = "sum")
plt.xlabel("scattering angle [deg.]", fontsize=16)
plt.ylabel("real", fontsize=16)
plt.legend(loc = 'lower center')

plt.subplot(122)
#plt.yscale('log')
plt.plot(theta, f1_i, color = 'black', linewidth=2.0, label = "analytical")
plt.plot(theta[0], f2_i[0], 'ro', linewidth=2.0, color = 'black', label = "sum")
for i in range(1, 18):
    plt.plot(theta[10 * i + 1], f2_i[10 * i + 1], 'ro', \
         linewidth=2.0, color = 'black')
plt.plot(theta[180], f2_i[180], 'ro', linewidth=2.0, color = 'black')
#plt.plot(theta, f2_i, color = 'black', linestyle = 'dashed', linewidth=2.0, \
#      label = "sum")
plt.xlabel("scattering angle [deg.]", fontsize=16)
plt.ylabel("imag", fontsize=16)
plt.legend(loc = 'upper center')
plt.show()
```

```
#          Program "bisphere_it_trans_input.py"
#


print()
print('   --- Input scattering configuration for acoustic plane ...')
print('        ... wave scattering on arbitrarily oriented ---')
print('                    ... bispheres! ---')
print()


import os
os.system("del input_data_bisphere_it_trans.txt")
os.system("type nul > input_data_bisphere_it_trans.txt")
fobj = open("input_data_bisphere_it_trans.txt", "w")


LS = int(input(' 0. iteration (0), 1. iteratio (1), or 2. iteration (2)? '))
print()
para0 = str(input('centered soft (s), hard (h) or penetrable (p) sphere? '))
para1 = str(input('shifted soft (s), hard (h) or penetrable (p) sphere? '))


if (para0 == 'p' and (para1 == 's' or para1 == 'h')):
    k_p_k_0 = float(input('ratio of wave numbers of centered sphere k_p0/k: '))
    rho_p_0 = float(input('density of the material inside the centered sphere \
    rho_p0: '))
elif (para1 == 'p' and (para0 == 's' or para0 == 'h')):
    k_p_k_1 = float(input('ratio of wave numbers of shifted sphere k_p1/k: '))
    rho_p_1 = float(input('density of the material inside the shifted sphere \
    rho_p1: '))
elif (para0 == 'p' and para1 == 'p'):
    k_p_k_0 = float(input('ratio of wave numbers of centered sphere k_p0/k: '))
    rho_p_0 = float(input('density of the material inside the centered sphere \
    rho_p0: '))
    k_p_k_1 = float(input('ratio of wave numbers of shifted sphere k_p1/k: '))
    rho_p_1 = float(input('density of the material inside the shifted sphere \
    rho_p1: '))
else:
    k_p_k_0 = 1.0
    rho_p_0 = 1.0
    k_p_k_1 = 1.0
    rho_p_1 = 1.0
print()
a0 = float(input('centered sphere with radius a0 [mm]: '))
a1 = float(input('shifted sphere with radius a1 [mm]: '))
beta_0 = float(input('size parameter of centered sphere beta_0: '))
```

```python
print()
b = float(input('distance between the centers of the spheres b [mm]: '))
alpha = float(input('Eulerian angle  alpha [degree]: '))
theta_p = float(input('Eulerian angle theta_p [degree]: '))
print()
ncut = int(input('truncation parameter ... ncut: '))
print()
plot_para = str(input('lin-lin (ll), or lin-log (lg) plot? '))
pl_pm1 = str(input("additional plot of Mie results (y/n)? "))


s00 = str(LS)
s00 = "LS = " + s00 + "\n"
fobj.write(s00)


s0 = "para0 = " + para0 + "\n"
fobj.write(s0)


s1 = "para1 = " + para1 + "\n"
fobj.write(s1)


s2 = str(k_p_k_0)
s2 = "k_p_k_0 = " + s2 + "\n"
fobj.write(s2)


s3 = str(rho_p_0)
s3 = "rho_p_0 = " + s3 + "\n"
fobj.write(s3)


s4 = str(k_p_k_1)
s4 = "k_p_k_1 = " + s4 + "\n"
fobj.write(s4)


s5 = str(rho_p_1)
s5 = "rho_p_1 = " + s5 + "\n"
fobj.write(s5)


s6 = str(a0)
s6 = "a0 = " + s6 + "\n"
fobj.write(s6)


s7 = str(a1)
s7 = "a1 = " + s7 + "\n"
fobj.write(s7)
```

```
s8 = str(beta_0)
s8 = "beta_0 = " + s8 + "\n"
fobj.write(s8)

s9 = str(b)
s9 = "b = " + s9 + "\n"
fobj.write(s9)

s9a = str(alpha)
s9a = "alpha = " + s9a + "\n"
fobj.write(s9a)

s9b = str(theta_p)
s9b = "theta_p = " + s9b + "\n"
fobj.write(s9b)

s10 = str(ncut)
s10 = "ncut = " + s10 + "\n"
fobj.write(s10)

s11 = "plot_para = " + plot_para + "\n"
fobj.write(s11)

s12 = "pl_pm1 = " + pl_pm1
fobj.write(s12)

fobj.close()
```

```
#        Program "bisphere_it_trans.py"
#
# This program calculates the plane wave scattering behaviour (zero-, first-,
# and second-order iteration) on a bisphere that is arbitrarily oriented in
# the laboratory frame. This program is based on a translation of the
# laboratory frame into the body frame! The configuration is fixed by the
# distance "b", and the Eulerian angles "\alpha" and "\theta_p".
#

import numpy as np
import scipy as scp
import scipy.special as scs
import matplotlib.pyplot as plt
import basics as bas
import os

# Reading the scattering configuration from input file:

fobj = open("input_data_bisphere_it_trans.txt", "r")

z = []
for line in fobj:
    line = line.strip()
    arr = line.split("= ")
    wert = str(arr[1])
    z = z + [wert]
fobj.close()

LS = int(z[0])
para0 = z[1]
para1 = z[2]
k_p_k_0 = float(z[3])
rho_p_0 = float(z[4])
k_p_k_1 = float(z[5])
rho_p_1 = float(z[6])
a0 = float(z[7])
a1 = float(z[8])
beta_0 = float(z[9])
b = float(z[10])
alpha = float(z[11])
theta_p = float(z[12])
ncut = int(z[13])
plot_para = z[14]
```

```
pl_pm1 = z[15]

k = beta_0 / a0
kb = k * b
beta_1 = k * a1
zi = 0. + 1.0j
pf = - zi / k

theta_l = 360.0
theta_l1 = 361
theta = np.linspace(0.0, theta_l, theta_l1)
ctheta = np.cos(theta * scp.pi / 180.)
ctheta_p = np.cos(theta_p * scp.pi / 180.)
r_theta = theta * scp.pi / 180.

# Calculation of the T-matrices:

if (para0 == 's' and para1 == 's'):
    tm_0 = bas.tm_s(ncut, beta_0)
    tm_1 = bas.tm_s(ncut, beta_1)
elif (para0 == 's' and para1 == 'h'):
    tm_0 = bas.tm_s(ncut, beta_0)
    tm_1 = bas.tm_h(ncut, beta_1)
elif (para0 == 's' and para1 == 'p'):
    tm_0 = bas.tm_s(ncut, beta_0)
    tm_1 = bas.tm_p(ncut, beta_1, a1, k, k_p_k_1, rho_p_1)
elif (para0 == 'h' and para1 == 's'):
    tm_0 = bas.tm_h(ncut, beta_0)
    tm_1 = bas.tm_s(ncut, beta_1)
elif (para0 == 'h' and para1 == 'h'):
    tm_0 = bas.tm_h(ncut, beta_0)
    tm_1 = bas.tm_h(ncut, beta_1)
elif (para0 == 'h' and para1 == 'p'):
    tm_0 = bas.tm_h(ncut, beta_0)
    tm_1 = bas.tm_p(ncut, beta_1, a1, k, k_p_k_1, rho_p_1)
elif (para0 == 'p' and para1 == 's'):
    tm_0 = bas.tm_p(ncut, beta_0, a0, k, k_p_k_0, rho_p_0)
    tm_1 = bas.tm_s(ncut, beta_1)
elif (para0 == 'p' and para1 == 'h'):
    tm_0 = bas.tm_p(ncut, beta_0, a0, k, k_p_k_0, rho_p_0)
    tm_1 = bas.tm_h(ncut, beta_1)
else:
    tm_0 = bas.tm_p(ncut, beta_0, a0, k, k_p_k_0, rho_p_0)
```

```
   tm_1 = bas.tm_p(ncut, beta_1, a1, k, k_p_k_1, rho_p_1)
t_o = tm_0
t_o1 = tm_1


# Calculation of the separation matrix:

sm_l_i_j = []
for l in range(0, ncut + 1):
   sm_i_j = []
   for i in range(0, ncut + 1):
      sm_j = []
      for j in range(0, ncut + 1):
         sm_j = sm_j + [bas.SM(l, i, j, kb)]
      sm_i_j = sm_i_j + [sm_j]
   sm_l_i_j = sm_l_i_j + [sm_i_j]


# Mie coefficients of the centered and shifted sphere:

c_m0_mie = []
q_m0_mie = []
for m0 in range(0, ncut + 1):
   d_m0 = np.sqrt(4. * np.pi * (2 * m0 + 1)) * zi**m0
   c_m0_mie = c_m0_mie + [t_o[m0] * d_m0]
   q_m0_mie = q_m0_mie + [np.exp(zi * kb * ctheta_p) * t_o1[m0] *\
              d_m0]


# Calculation of the differential scattering cross-section "dscross"
# as a function of the scattering angle "theta" in the interval [0, 2 \pi]
# in steps of 1 degree: only Mie theory for intercomparison purposes!

dscross_0 = []
dscross_1 = []
for i in range(0, theta_l1):
   psi_0_0 = 0.0
   psi_1_0 = 0.0

   for m0 in range(0, ncut + 1):
      Y_0m0 = scs.sph_harm(0,m0,0.,r_theta[i])
      psi_0_0 = psi_0_0 + c_m0_mie[m0] * (-zi)**m0 * Y_0m0
   psi_0_0 = pf * psi_0_0
   dscross_0 = dscross_0 + [psi_0_0 * np.conj(psi_0_0)]

   for m1 in range(0, ncut + 1):
```

```python
        Y_0m1 = scs.sph_harm(0,m1,0.,r_theta[i])
        psi_1_0 = psi_1_0 + q_m0_mie[m1] * (-zi)**m1 * Y_0m1
    psi_1_0 = pf * psi_1_0
    dscross_1 = dscross_1 + [psi_1_0 * np.conj(psi_1_0)]


# Iterative solutions in each local system:

c_l0_m0_0 = []
q_l1_n1_0 = []
for l0 in range(-ncut, ncut + 1):
    c_m0_1 = []
    q_n1_1 = []
    for m0 in range(np.abs(l0), ncut + 1):
        c_m0_0 = (c_m0_mie[m0] if l0 == 0 else 0.0)
        q_n1_0 = (q_m0_mie[m0] if l0 == 0 else 0.0)
        c_m0_1 = c_m0_1 + [c_m0_0]
        q_n1_1 = q_n1_1 + [q_n1_0]
    c_l0_m0_0 = c_l0_m0_0 + [c_m0_1]
    q_l1_n1_0 = q_l1_n1_0 + [q_n1_1]

c_l0_m0_1 = bas.Q12_tl_it1(ncut, alpha, theta_p, sm_l_i_j, t_o, q_l1_n1_0)
q_l1_n1_1 = bas.C21_tl_it1(ncut, alpha, theta_p, sm_l_i_j, t_o1, c_l0_m0_0)

c_it_1 = []
for l in range(-ncut, ncut + 1):
    v1 = c_l0_m0_0[l + ncut]
    v2 = c_l0_m0_1[l + ncut]
    v_it_1 = np.array(v1) - np.array(v2)
    c_it_1 = c_it_1 + [v_it_1]
q_it_1 = []
for l in range(-ncut, ncut + 1):
    w1 = q_l1_n1_0[l + ncut]
    w2 = q_l1_n1_1[l + ncut]
    w_it_1 = np.array(w1) - np.array(w2)
    q_it_1 = q_it_1 + [w_it_1]

cc1 = bas.C21_tl_it1(ncut, alpha, theta_p, sm_l_i_j, t_o1, c_it_1)
qq1 = bas.Q12_tl_it1(ncut, alpha, theta_p, sm_l_i_j, t_o, q_it_1)

c_l0_m0_2 = bas.Q12_tl_it1(ncut, alpha, theta_p, sm_l_i_j, t_o, cc1)
q_l1_n1_2 = bas.C21_tl_it1(ncut, alpha, theta_p, sm_l_i_j, t_o1, qq1)

c_final = []
```

```python
for l0 in range(-ncut, ncut + 1):
    v1 = c_l0_m0_0[l0 + ncut]
    v2 = c_l0_m0_1[l0 + ncut]
    v3 = c_l0_m0_2[l0 + ncut]
    if LS == 0:
        v = v1
    elif LS == 1:
        v = np.array(v1) - np.array(v2)
    else:
        v = np.array(v1) - np.array(v2) + np.array(v3)
    c_final = c_final + [v]

q_final = []
for l1 in range(-ncut, ncut + 1):
    w1 = q_l1_n1_0[l1 + ncut]
    w2 = q_l1_n1_1[l1 + ncut]
    w3 = q_l1_n1_2[l1 + ncut]
    if LS == 0:
        w = w1
    elif LS == 1:
        w = np.array(w1) - np.array(w2)
    else:
        w = np.array(w1) - np.array(w2) + np.array(w3)
    q_final = q_final + [w]

# Calculation of the differential scattering cross-section "dscross_g"
# as a function of the scattering angle "theta" in the interval [0, 2 \pi]
# in steps of 1.0 degree (in the laboratory frame):

psi_s_g = []
dscross_g = []
for i in range(0, theta_l1):
    c_k = kb * (np.cos(r_theta[i]) * ctheta_p + np.sin(r_theta[i]) * \
    np.sin(theta_p * scp.pi / 180.) * np.cos(alpha *scp.pi / 180.))
    psi_s = 0.0
    for l0 in range(-ncut, ncut + 1):
        c_l0 = c_final[l0 + ncut]
        q_l0 = q_final[l0 + ncut]
        for m0 in range(np.abs(l0), ncut + 1):
            m0z = m0 - np.abs(l0)
            pref = (-zi)**m0
            if i <= 180:
                Y_l0_m0 = scs.sph_harm(l0,m0,0.,r_theta[i])
```

```
        else:
            Y_l0_m0 = scs.sph_harm(l0,m0,np.pi,r_theta[360 - i])
            psi_s = psi_s + (c_l0[m0z] + np.exp(-zi * c_k) * q_l0[m0z]) * \
                    pref * Y_l0_m0
    psi_s = pf * psi_s
    psi_s_g = psi_s_g + [psi_s]
    dscross_g = dscross_g + [psi_s * np.conj(psi_s)]
dscross_g = np.real(dscross_g)
dscross_0 = np.real(dscross_0)
dscross_1 = np.real(dscross_1)
dscross_zw = 2. * (dscross_0 + dscross_1)
dscross_zw1 = dscross_0 + dscross_1   # für a1 << a0 !

# Generation of the result file:

os.system("del dscross_bisphere_it_trans.txt")
os.system("type nul > dscross_bisphere_it_trans.txt")
fobj1 = open("dscross_bisphere_it_trans.txt", "w")
nl = "\n"
for i in range(0, theta_l1):
    a1 = str(theta[i])
    a2 = " = "
    a3 = str(dscross_g[i]) + nl
    a = a1 + a2 + a3
    fobj1.write(a)
fobj1.close()

# Calculation of the total scattering cross-section "scat_tot" by use
# of the optical theorem:

print()
print()
print('Results: ')
w = np.imag(psi_s_g[0])
scat_tot = 4 * scp.pi * w / k
print()
print("total scattering cross-section: scat_tot = ", scat_tot)
print()

# Plot of the results:

if plot_para == 'lg':
    plt.yscale('log')
```

```python
    if pl_pm1 == 'y':
        plt.plot(theta, dscross_zw, '--', color = 'black', linewidth=2.0, \
            label = "2 * (Mie_a0 + Mie_a1)")
        plt.plot(theta, dscross_g, color = 'black', linewidth=2.0, \
            label = "bisphere")
    else:
        plt.plot(theta, dscross_g, color = 'black', linewidth=2.0, \
            label = "bisphere")
else:
    if pl_pm1 == 'y':
        plt.plot(theta, dscross_zw, '--', color = 'black', linewidth=2.0, \
            label = "2 * (Mie_a0 + Mie_a1)")
        plt.plot(theta, dscross_g, color = 'black', linewidth=2.0, \
            label = "bisphere")
    else:
        plt.plot(theta, dscross_g, color = 'black', linewidth=2.0, \
            label = "bisphere")
plt.xlabel("scattering angle [deg.]", fontsize=16)
plt.ylabel("diff. scat. cross-sect.", fontsize=16)
plt.legend()
#plt.legend(loc = 'lower right')
plt.legend(loc = 'upper center')
#plt.legend(loc = 'upper left')
plt.show()
```

Printed in the United States
By Bookmasters